NATIONAL ACADEMIES Sciences Engineering Medicine

NATIONAL ACADEMIES PRESS
Washington, DC

# Progress and Priorities in Ocean Drilling

## In Search of Earth's Past and Future

2025–2035 Decadal Survey of Ocean Sciences for the National Science Foundation

Ocean Studies Board

Division on Earth and Life Studies

Consensus Study Report

**NATIONAL ACADEMIES PRESS 500 Fifth Street NW Washington, DC 20001**

This activity was supported by a contract between the National Academy of Sciences and the National Science Foundation. Any opinions, findings, conclusions, or recommendations expressed in this publication do not necessarily reflect the views of any organization or agency that provided support for the project.

International Standard Book Number-13: 978-0-309-71338-2
International Standard Book Number-10: 0-309-71338-2
Digital Object Identifier: https://doi.org/10.17226/27414
Library of Congress Control Number: 2024940946

This publication is available from the National Academies Press, 500 Fifth Street, NW, Keck 360, Washington, DC 20001; (800) 624-6242 or (202) 334-3313; http://www.nap.edu.



Printed in the United States of America.

Suggested citation: National Academies of Sciences, Engineering, and Medicine. 2024. *Progress and Priorities in Ocean Drilling: In Search of Earth's Past and Present*. Washington, DC: The National Academies Press. https://doi.org/10.17226/27414.

The **National Academy of Sciences** was established in 1863 by an Act of Congress, signed by President Lincoln, as a private, nongovernmental institution to advise the nation on issues related to science and technology. Members are elected by their peers for outstanding contributions to research. Dr. Marcia McNutt is president.

The **National Academy of Engineering** was established in 1964 under the charter of the National Academy of Sciences to bring the practices of engineering to advising the nation. Members are elected by their peers for extraordinary contributions to engineering. Dr. John L. Anderson is president.

The **National Academy of Medicine** (formerly the Institute of Medicine) was established in 1970 under the charter of the National Academy of Sciences to advise the nation on medical and health issues. Members are elected by their peers for distinguished contributions to medicine and health. Dr. Victor J. Dzau is president.

The three Academies work together as the **National Academies of Sciences, Engineering, and Medicine** to provide independent, objective analysis and advice to the nation and conduct other activities to solve complex problems and inform public policy decisions. The National Academies also encourage education and research, recognize outstanding contributions to knowledge, and increase public understanding in matters of science, engineering, and medicine.

Learn more about the National Academies of Sciences, Engineering, and Medicine at **www.nationalacademies.org**.

**2025–2035 DECADAL SURVEY OF OCEAN SCIENCES**
**FOR THE NATIONAL SCIENCE FOUNDATION**

**TUBA OZKAN-HALLER,** *Co-Chair*, Oregon State University
**JAMES (JIM) YODER,** *Co-Chair,* Woods Hole Oceanographic Institution (emeritus)
**LIHINI ALUWIHARE,** Scripps Institution of Oceanography, University of California, San Diego
**MONA BEHL,** University of Georgia
**MARK BEHN,** Boston College
**BRAD deYOUNG,** Canadian Integrated Ocean Observing System
**CARLOS GARCIA-QUIJANO,** University of Rhode Island
**PETER GIRGUIS,** Harvard University
**LEILA J. HAMDAN,** University of Southern Mississippi
**MARCIA ISAKSON,** Applied Research Laboratories, University of Texas at Austin
**JASON LINK,** National Oceanic and Atmospheric Administration
**ALLISON MILLER,** Schmidt Ocean Institute
**S. BRADLEY MORAN,** University of Alaska Fairbanks
**RICHARD W. MURRAY,** Woods Hole Oceanographic Institution
**STEPHEN R. PALUMBI,** Stanford University
**ELLA (JOSIE) QUINTRELL,** Integrated Ocean Observing System (retired)
**YOSHIMI (SHIMI) M. RII,** Hawai'i Institute of Marine Biology, University of Hawai'i
**KRISTEN ST. JOHN,** James Madison University
**SAMUEL KERSEY STURDIVANT,** INSPIRE Environmental
**AJIT SUBRAMANIAM,** Columbia University
**MAYA TOLSTOY,** University of Washington College of the Environment
**SHANNON VALLEY,** Vistant
**JAMES ZACHOS,** University of California, Santa Cruz

*Study Staff*

**KELLY OSKVIG,** Senior Program Officer
**LEIGHANN MARTIN,** Associate Program Officer (until January 2024)
**ZOE ALEXANDER,** Senior Program Assistant
**ERIK YANISKO,** Program Assistant (until January 2024)

*Sponsor*

**NATIONAL SCIENCE FOUNDATION**

## OCEAN STUDIES BOARD

# Reviewers

This consensus study report was reviewed in draft form by individuals chosen for their diverse perspectives and technical expertise. The purpose of this independent review is to provide candid and critical comments that will assist the National Academies of Sciences, Engineering, and Medicine in making each published report as sound as possible and to ensure that it meets the institutional standards for quality, objectivity, evidence, and responsiveness to the study charge. The review comments and draft manuscript remain confidential to protect the integrity of the deliberative process.

We thank the following individuals for their review of this report:

**ANDREA AHRENS,** Stantec
**BARBARA BEKINS,** U.S. Geological Survey (NAE)
**JULIE HUBER,** Woods Hole Oceanographic Institution
**BO BARKER JORGENSEN,** Aarhus University (NAS)
**ADRIAN LAM,** Binghampton University
**WILLIAM MILLER,** University of Georgia
**MAUREEN RAYMO**, Columbia University (NAS)
**DAMIAN SAFFER,** University of Texas at Austin
**JOHN SHERVAIS,** Utah State University

Although the reviewers listed above provided many constructive comments and suggestions, they were not asked to endorse the conclusions or recommendations of this report nor did they see the final draft before its release. The review of this report was overseen by **ROBERT DUCE,** Texas A&M University, and **LARRY MAYER (NAE),** University of New Hampshire. They were responsible for making certain that an independent examination of this report was carried out in accordance with the standards of the National Academies and that all review comments were carefully considered. Responsibility for the final content rests entirely with the authoring committee and the National Academies.

# Acknowledgments

The committee thanks the following individuals for their contributions during the study process, especially for enriching and informing the discussions at the open session meetings of the committee: James Allen (National Science Foundation [NSF]), Jennifer Biddle (University of Delaware), Donna Blackman (University of California, Santa Cruz), Stefanie Brachfeld (Montclair State University), Carl Brenner (U.S. Science Support Program), Steven D'Hondt (University of Rhode Island), Patrick Fulton (Cornell University), Sean Gulick (University of Texas), David Hodell (University of Cambridge), Celli Hull (Yale University), Minoru Ikehara (Kochi University), Fumio Inagaki (Japanese Agency for Marine-Earth Science and Technology), Kevin Johnson (NSF), Brandi Kiel Reese (University of South Alabama), Anthony Koppers (Oregon State University), Larry Krissek (The Ohio State University), Jessica Labonté (Texas A&M University, Galveston), Adriane Lam (Binghamton University), Chris Lowery (University of Texas), Mitch Malone (Texas A&M University), Kathie Marsaglia (California State University, Northridge), Robert McKay (Victoria University of Wellington), Lisa McNeil (Southampton University), Charna Meth (International Ocean Discovery Program Science Support Office), Heiko Palike (University of Bremen), Becky Robinson (University of Rhode Island), Yair Rosenthal (Rutgers University), Demian Saffer (University of Texas), Daniel Sigman (Princeton University), David Smith (University of Rhode Island), Chijun Sun (National Center for Atmospheric Research), Jason Sylvan (Texas A&M University), Allyson Tessin (Kent State University), Masako Tominaga (Woods Hole Oceanographic Institution), Maureen Walczak (Oregon State University), Allen Walker (NSF's Technology, Innovation, and Partnership), Shelby Walker (NSF), and Trevor Williams (Texas A&M University). Their input was critical to the completion of the committee's work.

The committee would also like to thank our primary contact at NSF's Division of Ocean Sciences, Jim McManus, for his efforts in developing and sponsoring this study and for providing important documents and support upon the committee's request.

# Contents

**APPENDIXES**

# Boxes, Figures, and Tables

## BOXES



## FIGURES

## TABLES

# Acronyms and Abbreviations

| | |
|---|---|
| AMOC | Atlantic meridional overturning circulation |
| ARF | U.S. Academic Research Fleet |
| DSDP | Deep Sea Drilling Project |
| DSOS | Decadal Survey of Ocean Sciences |
| ECORD | European Consortium for Ocean Research Drilling |
| eODP | Extending Ocean Drilling Pursuits |
| EPO | expedition project manager |
| FAIR | findable, accessible, interoperable, reusable |
| GMSL | global mean sea level |
| GNSS | global navigation satellite system |
| HLAPC | half-length advanced piston corer |
| HRT | hydraulic release tool |
| ICDP | International Continental Scientific Drilling Program |
| IODP-1 | Integrated Ocean Drilling Program |
| IODP-2 | International Ocean Discovery Program |
| JRSO | *JOIDES Resolution* Science Operator |
| kyr | thousand years |
| LEAP | legacy asset project |
| Ma | million years ago |
| MARUM | Center for Marine Environmental Sciences (German) |

| | |
|---|---|
| MSP | mission-specific platform |
| myr | million years |
| NanTroSEIZE | Nankai Trough Seismogenic Zone Experiment |
| NSF | National Science Foundation |
| OCAP | Ocean Climate Action Plan |
| ODP | Ocean Drilling Program |
| OMZ | oxygen minimum zone |
| OOI | Ocean Observatories Initiative |
| PETM | Paleocene–Eocene thermal maximum |
| PMIP | Paleoclimate Model Intercomparison Project |
| ppm | parts per million |
| Pg | petagram |
| *SMR* | *Science Mission Requirements* (report of the U.S. Science Support Program) |
| STEM | science, technology, engineering, and mathematics |
| SZ4D | Subduction Zones in four Dimensions |
| UNOLS | University-National Oceanographic Laboratory System |
| WAIS | West Antarctic Ice Sheet |

# Summary

Research supported by scientific ocean drilling has fundamentally transformed the understanding of the planet, with key scientific contributions to the discovery and understanding of plate tectonics; the formation and destruction of ocean crust and how these processes generate geohazards; the reconstruction of extreme greenhouse and icehouse climates that existed during the past 100 million years; the identification of major extinctions; and the discovery of a diverse community of microbes living in ocean sediments, rocks, and fluids below the seafloor.

Scientific ocean drilling is now at a critical juncture. The U.S. dedicated drilling vessel for deep-sea research, the *JOIDES Resolution*,[1] has been the workhorse for collaborative international scientific ocean drilling for decades. The contract to operate the *JOIDES Resolution* has not been renewed, and operations will end in 2024. Currently, there is no plan in place for a future dedicated U.S. drilling vessel. Meanwhile, U.S. scientific ocean drilling's international partners in Europe and Japan are jointly moving forward with plans for a new program phase, with berths on contracted vessels available to contributing member countries. The United States has not joined this consortium (branded as the International Ocean Drilling Program-3 [IODP$^3$]). Additionally, China is developing a new scientific ocean drilling program independently. Thus, the landscape for scientific ocean drilling will change after 2024.

With the absence of a dedicated drilling vessel supported by the United States, the capacity for future scientific ocean drilling for the United States and its present international partners will likely be reduced to approximately 10 percent of its current capacity. Without new infrastructure or sampling investments, participation of U.S. scientists on expeditions will become limited, and access to new ocean drilling samples and data will be curtailed. These conditions will impact progress on globally vital and urgent research. This includes research on pressing topics such as ground truthing predictive models of climate and ocean ecosystem changes, monitoring and assessing tectonically generated geologic hazards, and evaluating the potential for carbon sequestration in the ocean crust.

There are high-priority science questions, with potential to yield societal benefits, that can be addressed only with ocean drilling research. In order to advance that research, new approaches addressing resources, infrastructure, and capacity need to be considered.

[1] "JOIDES" is an acronym for the Joint Oceanographic Institutions for Deep Earth Sampling (see https://joidesresolution.org/about-the-jr).

## PLANNING FOR THE FUTURE PROGRAM

The United States has led the international scientific ocean drilling community since 1968. The *JOIDES Resolution*, funded by the National Science Foundation and its operational team, provides essential leadership. The *JOIDES Resolution* and its predecessor vessel have operational capabilities that are unique for scientific ocean drilling—the ability to operate in deep water and collect continuous samples (cores and associated materials) to depths exceeding 1 kilometer below the seafloor under a wide range of sea conditions. Since beginning operations in 1985, the *JOIDES Resolution* has collected 95 percent of the total core length for international scientific ocean drilling, with free and open access to samples and data from these cores available to the international Earth and ocean science community. U.S. scientists have also led the scientific ocean drilling community in the conception and design of drilling projects and in the dissemination of research results (e.g., publications) and collaborations.

Owing to rising costs that now far exceed the funding designated to operate the *JOIDES Resolution*, NSF will terminate support for the vessel in 2024. As possible next steps, NSF is considering a U.S. mission-specific platform (MSP) program, and alternative drilling and coring options may be possible in the future; however, these options will not achieve the vital and urgent objectives identified in this report. Furthermore, no funding mechanism has been identified for building or operating a vessel with capabilities similar to those of the *JOIDES Resolution* that could address these science objectives.

In preparation for the impending retirement of the *JOIDES Resolution*, and at the request of NSF, the scientific ocean drilling community has taken steps to envision and plan for the next phase of the drilling program. This effort has included international and U.S. identification of priority science areas and U.S. conceptualization of drilling vessel requirements. Additionally, NSF asked the National Academies of Sciences, Engineering, and Medicine's Committee on the 2025–2035 Decadal Survey of Ocean Sciences for the National Science Foundation to produce and publish this consensus study, serving as an interim report for a more encompassing decadal survey to come. The interim report provides a timely and broad perspective on critical research and infrastructure needed to answer the most compelling research questions that can be advanced only with scientific ocean drilling. The committee's statement of task for this interim report is included in Box S.1.

**BOX S.1**
**Interim Report Statement of Task**

The Committee will produce an interim report to provide advice to NSF OCE on the resources and infrastructure available to address high priority research questions requiring scientific ocean drilling. The interim report will cover the following:

1. Based on previous reports, assess progress on addressing high priority science questions that require scientific ocean drilling and identify new, if any, equally compelling science questions that would also require scientific ocean drilling.

2. Of the unanswered scientific questions, which could be addressed using existing scientific drilling assets including sediment or rock core archives and existing platforms, and which questions would require new infrastructure or sampling investments?

## PROGRESS OVER THE LAST DECADE

Since the scientific ocean drilling program commenced in 1968, scientific ocean drilling expeditions and research have fundamentally transformed the understanding of the planet by revealing the critical features of Earth's dynamic history, processes, and structure, including the solid Earth (i.e., upper mantle/crust), ocean, atmosphere, and ecosystems. The list of scientific ocean drilling–related achievements is extensive, represented in large part by the number of publications and, until recently, by the sustained support for drilling by the international Earth sciences community.

Over roughly the last decade, the scientific ocean drilling program has operated within the funding phase branded as the International Ocean Discovery Program (IODP-2). From 2014 to 2023, IODP-2 completed 57 expeditions: 46 using the *JOIDES Resolution*, 5 using the *Chikyu*, and 6 using MSPs. This extensive use of the *JOIDES Resolution,* compared with the other IODP-2 components, illustrates the value and productivity of the *JOIDES Resolution*. While research from the current phase of the program is still in progress, the findings from several expeditions have already yielded important scientific insights, as illustrated by the sample accomplishments listed below:

***Climate and ocean change*: Reading the past, informing the future**

- Documented the influence of ice sheet dynamics on the magnitude of sea level change.
- Quantified data on ocean circulation and climate sensitivity to changing greenhouse gas levels.
- Progressed understanding of regional monsoon precipitation.
- Identified physical and biogeochemical changes that affect ecosystems and climate.

***Earth connections*: Deep processes and their impact on Earth's surface environment**

- Developed new insights and models for chemical and fluid exchanges between ocean crust and seawater.
- Fulfilled a 60-year goal of scientific ocean drilling by successfully drilling into and recovering upper mantle rock.
- Elucidated the processes by which ocean crustal architecture is created and modified from rifting to seafloor spreading.
- Increased understanding of how mantle melting processes evolve during and after subduction initiation.

***Biosphere frontiers*: Deep life and environmental forcing of evolution**

- Revealed global diversity of microbial communities in subseafloor environments (e.g., sediments, rocks, fluids).
- Made pioneering observations about microbial life in extreme environments, including the limits of life. Documented ecosystem responses to major environmental shifts.
- Revealed how marine microbial and planktonic communities respond to pronounced changes in climate, including events that led to mass extinctions.

***Earth in motion*: Processes and hazards on human timescales**

- Documented new carbon cycling links between Earth's surface and its deeper interior along plate boundaries, making connections to climate change.
- Advanced characterization of fluid flow in a range of environmental settings using borehole instruments and core records.
- Made major progress in deep drilling, sampling, and borehole instrumentation of plate boundaries to understand a range of fault types, geologic properties, and motions leading to earthquakes.
- Provided new recognition of climatically linked submarine landslides.

It should be noted that the scientific ocean drilling program has not conducted a formal evaluation of the scientific progress made and would benefit from developing and executing such an evaluation to assess progress and communicate the program's achievements and value.

## FUTURE SCIENTIFIC OCEAN DRILLING PRIORITIES

Funding for scientific research is not unlimited; forward-looking prioritization is needed to guide investments in research, infrastructure, and workforce development. Important research that can be advanced only using scientific ocean drilling is identified here in recognition of the serious nature of regional and global change, the risks of geologic hazards, and the research areas of greatest societal impact. With this context, the committee defined two prioritization categories to provide guidance to NSF: *vital science* and *urgent science*.

*Vital science* encompasses compelling, high-priority research that has the potential to transform scientific knowledge of the interconnected Earth system and the critical role of the ocean in that system. Vital scientific research can lead to paradigm shifts in understanding, potentially opening new doors to research and technology innovations that can benefit humanity with direct societal relevance.

*Urgent science* is vital research that is time sensitive and has immediate societal relevance given the emerging challenges at regional to global scales. Research on urgent science questions needs to be done now to understand changes or new circumstances that can inform predictive models and decision making and may be related to tipping point vulnerabilities. It implies that immediate action is needed and is thus a higher priority than vital science.

The committee identified five high-priority research areas (Figure S.1, described in the text below) that continue to require scientific ocean drilling in order to advance understanding. Each of these high-priority research areas are vital; some are additionally classified as urgent because of their direct societal benefits, such as predicting the collapse of deep-sea currents that regulate global temperatures or predicting future geohazards.

This interim report also identifies first-order infrastructure parameters required to accomplish these vital and urgent ocean drilling science research priorities.

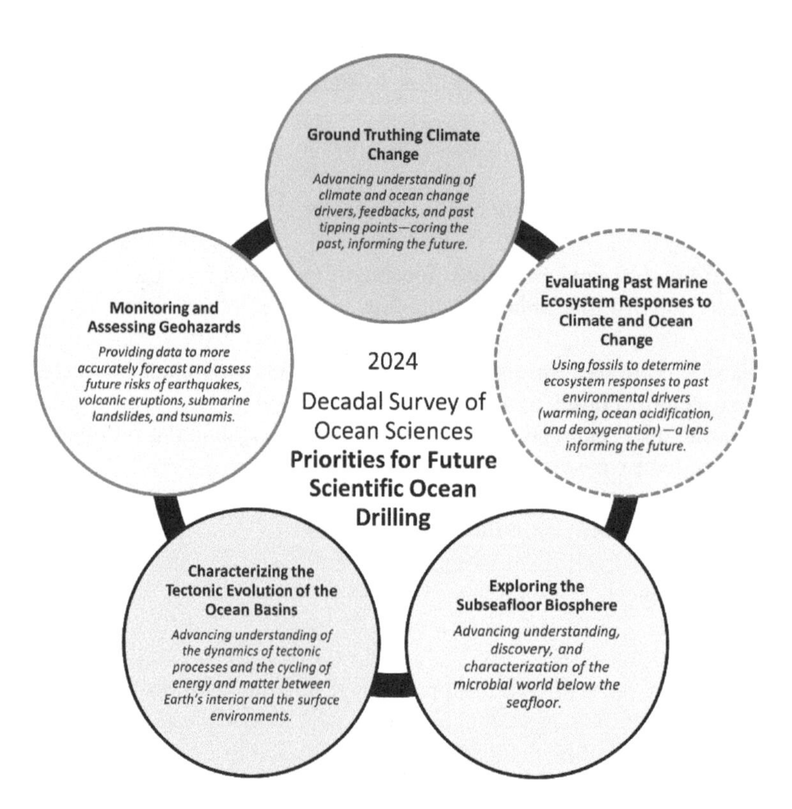


**FIGURE S.1** Priorities for future scientific ocean drilling. NOTE: All five priority areas are considered vital; those outlined in red are also deemed urgent.

### Ground Truthing Climate Change

*Advancing understanding of climate and ocean change drivers, feedbacks, and past tipping points—coring the past, informing the future.*

*Ground truthing climate change* is deemed both vital and urgent because records recovered through scientific ocean drilling are important past analogs for modern and near-future challenges of rapid global warming, sea level rise, and widespread ocean acidification and deoxygenation. Data from these records are useful in informing predictive models today because they contain paleoclimate proxies of climate and ocean variables. The proxy data collected and measured serve essentially as surrogates, or indirect indicators, of past changes in temperature, ice volume, and ocean chemistry, among others. Direct observations of global climate from less than a century ago provide too little data to adequately assess the ability of advanced models to accurately simulate Earth's climate at greenhouse gas levels significantly higher (or lower) than present. Additionally, long, continuous, and high-resolution paleoclimatic and paleoceanographic sedimentary records from the subseafloor are useful to constrain the processes that regulate or destabilize feedbacks in Earth's climate system and to examine the geological record of past tipping points (critical points that, when exceeded, will lead to large and often irreversible change); transient climate states; and the dynamics of ice, ocean, and atmosphere interactions in past periods of elevated temperatures.

**Future Research:** Additional observations obtainable only by scientific ocean drilling are required to assess the skill of climate models to replicate greenhouse gas–forced switches (i.e., tipping points) over geological timescales in temperatures, ice sheet dynamics, sea level, and ocean circulation and to constrain the role of feedbacks (physical or biogeochemical responses that amplify or dampen perturbations). To constrain Earth climate sensitivity to high greenhouse gas levels, additional scientific drilling is required to fill data gaps for extreme warm intervals in climatically sensitive regions (e.g., the Arctic and equatorial oceans, and in a few cases, the midlatitudes). Similarly, to fully characterize the sensitivity of hydroclimates (including regional monsoons) to greenhouse gas forcing, records obtained for the Northern Hemisphere need to be complemented with records from the Southern Hemisphere.

### Evaluating Past Marine Ecosystem Responses to Climate and Ocean Change

*Using fossils to determine ecosystem responses to past environmental drivers (warming, ocean acidification, and deoxygenation)—a lens informing the future.*

*Evaluating past marine ecosystem responses to climate and ocean change* is vital and potentially urgent, especially given the importance of marine ecosystems as a source of seafood. Just as the past can help understand drivers of climate change, so too can it provide insights into marine ecosystem responses across multiple timescales and geographies. This topic necessarily goes hand in hand with *ground truthing climate change*. Records that can be recovered only through scientific ocean drilling can provide insight into past ecosystem responses to accelerated changes in climate and ocean. Such records provide a framework and foundation to situate modern studies of changing climate and ocean conditions, and provide a necessary long-term context for assessing impacts and feedbacks on ecosystem dynamics and food webs. Collectively, these data inform predictive models of future change.

**Future Research:** Additional scientific ocean drilling that prioritizes locations with limited records, such as the equatorial, midlatitude, and polar oceans and open ocean environments during past periods of extreme warmth, will allow paleobiologists to inform models of plankton ecosystem dynamics during past analog climate states (e.g., rapid warming). In addition, existing long-term paleo records can be further exploited for studies, capitalizing on the development of new databases and existing core samples to assess global marine ecosystem responses to climatic and oceanic shifts more fully.

### Monitoring and Assessing Geohazards

*Providing data to more accurately forecast and assess future risks of earthquakes, volcanic eruptions, submarine landslides, and tsunamis.*

Deep, subseafloor observatories, which can only be installed through scientific ocean drilling, are used for what is considered both vital and urgent research under the theme of *monitoring and assessing geohazards* (earthquakes, volcanic eruptions, submarine landslides, and tsunamis). These observatories are an order of magnitude more sensitive to fault slip than other real-time systems (e.g., seabed observations), allowing smaller events to be detected and improving the potential for future earthquake forecasting. The results of this research could have direct societal benefit in terms of preparing for and better mitigating future geohazard risks.

**Future Research:** Future studies of subduction systems, including borehole monitoring, will allow scientists to better understand different conditions that promote either seismogenic or stable (non-earthquake-producing) fault motion. These data will constrain numerical models of dynamic fault ruptures, earthquake cycles, and tsunami genesis to advance understanding of the conditions under which natural hazards occur and to create a more robust warning system.

### Exploring the Subseafloor Biosphere

*Advancing understanding, discovery, and characterization of the world of living microbes below the seafloor.*

*Exploring the subseafloor biosphere* is characterized as vital; although findings could have direct societal benefit, the research is exploratory in nature. Building on pioneering research by ocean drilling over the past decade in particular, those exploring subsurface microbial life are on the cusp of discoveries that are expected to transform scientific understanding of microbial activity in extreme environments. Subseafloor sediment and hard rock records are essential to understanding what makes the planet habitable, and where and how life originated and evolved. Understanding the limits of life requires knowledge of the complex exchanges of fluids and nutrients that occur between the subseafloor biosphere, Earth's crust, the ocean, and the atmosphere. Research has also shown that the subsurface biosphere—including the deeper subsurface environs—can have a pronounced impact on biogeochemical cycles. However, few studies to date have explicitly focused on biogeochemical and ecological coupling (e.g., the extent of community exchange between the subsurface and the overlying water column).

**Future Research:** Scientific ocean drilling is necessary to address key unanswered questions about the subseafloor biosphere and to advance understanding of the limits to life, as well as the way biological communities interact and move within the subsurface biosphere and how they are distributed across space and time. Such research has direct implications for understanding the potential for life in other areas of the solar system, the origins of life on Earth, and the integral building blocks of ecosystems that nurture the biological world.

### Characterizing the Tectonic Evolution of the Ocean Basins

*Advancing understanding of the dynamics of tectonic processes and the cycling of energy and matter between Earth's interior and surface environments.*

*Characterizing the tectonic evolution of the ocean basins* is vital, high-priority research. Sampling oceanic crust of different ages provides insight into Earth processes that govern the occurrence of earthquakes, tsunamis, and volcanoes and the global cycling of energy and matter that produces economic resources that have importance now (i.e., oil and gas) and that have potential to become an economic resource in the future (i.e., critical minerals).

**Future Research:** Only scientific ocean drilling can provide key constraints regarding the formation and evolution of oceanic crust and the upper mantle. The cycling of fluids through the subseafloor and corresponding chemical exchanges between the liquid and solid Earth have implications for processes with direct societal relevance, including the production of mineral resources, sequestration of atmospheric carbon dioxide, and origin of geohazards (including volcanic eruptions, earthquakes, and related tsunamis).

Although they differ in detail and nuance, the five priority areas identified by the committee align with the initiatives identified throughout the years by the scientific ocean drilling community; they also connect and respond directly to U.S. research priorities identified by the White House, by the scientific ocean drilling community, and by several previous National Academies studies (Table S.1).

**TABLE S.1** Connecting Scientific Ocean Drilling Priorities to U.S. National Priorities and Prior Study Recommendations to NSF

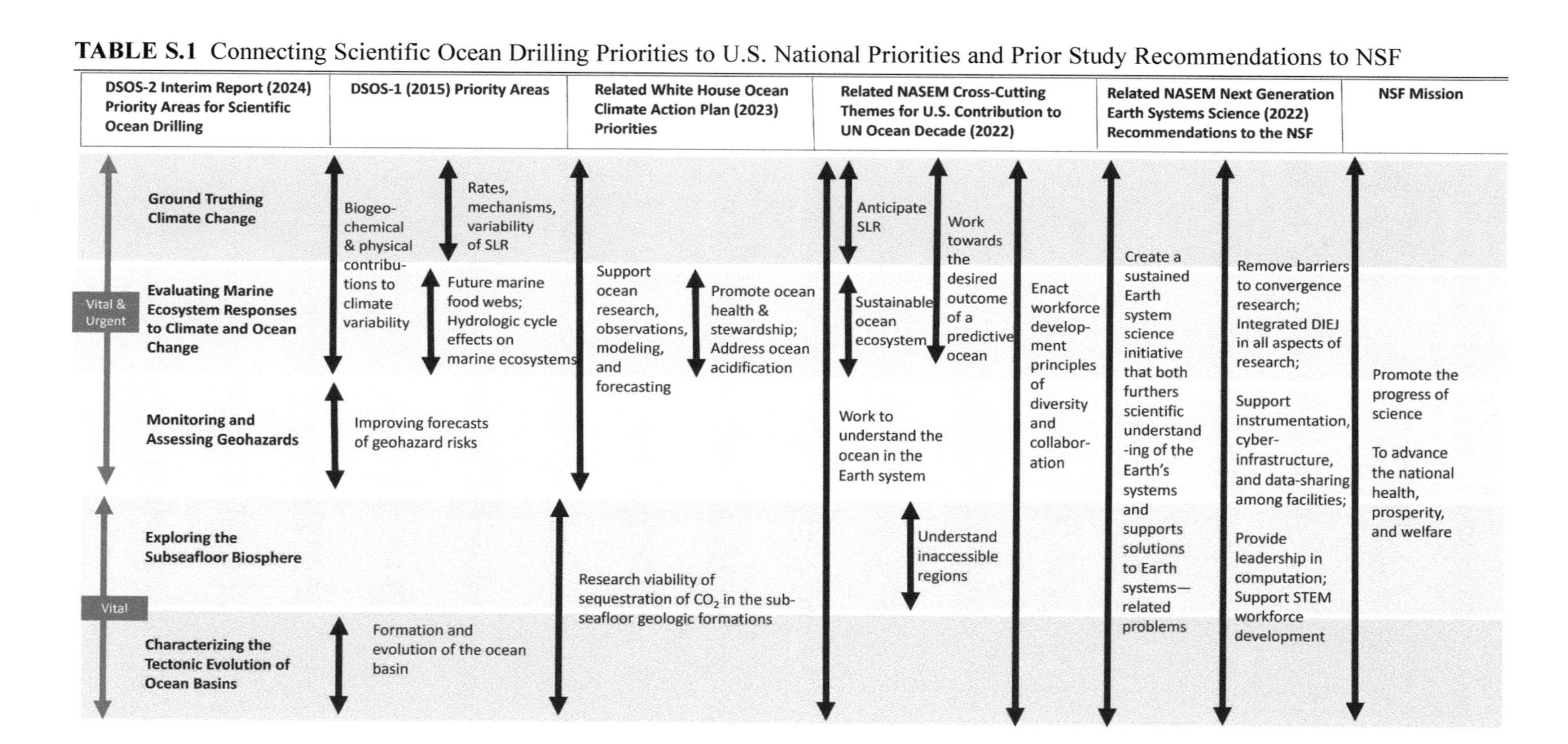


NOTES: DSOS = Decadal Survey of Ocean Sciences; DEIJ = diversity, equity, inclusion, and justice; NSF = National Science Foundation; SLR = sea level rise; STEM = science, technology, engineering, and mathematics.

## USING RESOURCES WISELY—LEGACY ASSETS

Opportunities exist to use available archived materials (i.e., legacy assets), including collected cores, data, and other samples, to accomplish groundbreaking and essential scientific research (Table S.2). However, not all scientific objectives will be met without collection of new cores (and associated data) and installation of new observatories. A holistic approach to understanding the high-priority scientific areas includes strategic use of existing archives along with targeted drilling for new records and access to the subseafloor.

Across scientific ocean drilling priorities, legacy assets can help address the following types of research and analysis:

- Community-driven, collaborative, multidisciplinary research.
- Big data analytics on a wide range of subseafloor standard measurements (e.g., physical properties, petrophysics/logging, paleomagnetic data).
- Large-scale "syntheses of science" studies (i.e., producing topical reviews) that integrate data across multiple expeditions/boreholes, addressing global or regional geographies and time intervals.
- Development and testing of new proxy methods (that are not dependent on ephemeral properties).
- Undergraduate and graduate education and training on materials and methods used in scientific ocean drilling research to engage early-career scientists and develop workforce-ready skills.

High-priority scientific ocean drilling research and analyses that *cannot* be advanced using legacy assets include:

- Sample-intensive, high-resolution studies using cores that have already been heavily sampled (Figure S.2).
- Real-time monitoring of fault motion using existing borehole instruments.
- Microbiological and biogeochemical studies requiring fresh samples/biomass.
- Comprehensive studies of igneous and metamorphic rocks, as there is very little repository material of these rock types.
- Studies of challenging rock types, such as those in fault zones, because there is little repository material of intervals from such locations (the exception being those samples from along the Japan margin).
- Analyses that depend on ephemeral properties (e.g., pore water, organic carbon).

**TABLE S.2** Available Assets Obtained from Select Scientific Ocean Drilling Programs* to Advance Some Vital and Urgent Research Priorities

| | |
|---|---|
| Cores | Approximately 150 km of collected core from all drilling platforms are stored in each of three locations: Gulf Coast (United States), Bremen (Germany), and Kochi (Japan), for a total of ~450 km. Approximately one-third of the total core length is appropriate for high-priority science. Cores of high scientific and societal interest are quickly depleted by use. Also, cores recovered in earlier phases of the program are now dried, and some are contaminated by mold (a natural consequence of long-term core storage), which makes obtaining chemical data from these older sediments difficult. |
| Data | Approximately 1,000,000 unique measurements per drilling expedition, and ~700 core images and ~700 X-ray images per km of each core exist. Continued support for data stewardship activities is critical to handle issues related to data quality, resolution of the time record, calibration, and combining datasets. |
| Microbiology Samples | Approximately 1,300 samples exist that are frozen for preservation for molecular analysis. However, freezing commonly limits usability in future analysis, and past experience has highlighted challenges in storage. For example, frozen samples are not suitable for determining microbial activity or rates of activity and are unsuitable for any potential cultivation work. |
| Instrumented Boreholes | Approximately 50 borehole active observatories exist, but few transmit data in real time and most require revisitation to install/reinstall apparatus and/or download data. Additionally, ~90 inactive borehole observatories are ready for reentry and reinstrumentation by a vessel, if determined practical and feasible. |

*Select scientific drilling programs include the Deep Sea Drilling Project, Ocean Drilling Program, International Ocean Discovery Programs 1 and 2.

**FIGURE S.2** Highly sampled core from an interval of high scientific interest. NOTE: Styrofoam spacers (white) replace samples that have already been used.
SOURCE: *JOIDES Resolution* Science Operator.

- Creation of high-temporal-resolution geochemical records obtained from carbonate and molecular fossil materials to reconstruct ocean conditions (e.g., temperature, salinity); this limitation is due to degradation of fossil material from acidic pore waters present in the legacy cores.
- Coordinated land–sea studies (e.g., continental drilling and ocean drilling studies), either because existing cores and other assets were not necessarily taken from locations best positioned for linking to adjacent continental records, or continental and ocean records are not yet available (e.g., not yet drilled).

A new approach to collaborative research has been proposed by the scientific ocean drilling community, with the first call for proposals for Legacy Asset Projects (LEAPs) issued in October 2023 by the IODP Science Support Office. LEAPs provide opportunities to maximize the use of already acquired material and data and foster discovery and innovation. However, many of the vital and urgent science priorities cannot be addressed with stored material (e.g., preservation of microbiology samples limits their usability) and the core materials in critical locations and intervals have been depleted by use yet remain in high demand (Figure S.2). A funded LEAP initiative would not replace the need for drilling capacity. An ideal scientific drilling program would include a robust LEAPs program combined with recovery of new subseafloor cores and installation of borehole observatories that address the five high-priority research areas.

While funding the use of legacy assets is important, it is equally important that the metadata and data collected, regardless of type or source, be findable, accessible, interoperable, reproducible (FAIR), and shared in a timely manner. Timely sharing is essential and could be incentivized and valued as much as publications of expedition outcomes. Crediting use-inspired data sharing, as one produces a publication, has the potential to ensure that the culture includes FAIR, responsible, ethical (e.g., CARE principles[2]), and timely data sharing, not just in the scientific ocean drilling program, but throughout the field of ocean science.

## INFRASTRUCTURE NEEDS

The committee identified key criteria, or parameters, for successful achievement of the scientific goals associated with each of the five high-priority areas for future scientific ocean drilling. Rather than recommend any specific path forward in terms of drilling infrastructure, key parameters necessary for successfully fulfilling the scientific themes categorized as vital and/or urgent are identified in the column headings in Table S.3. The parameters are not an exhaustive list but define the high-level screening parameters relevant to generating the data needed by scientists to address these themes, which are listed as rows in Table S.3.

[2] CARE stands for Collective benefit, Authority to control, Responsibility, Ethics (see https://www.gida-global.org/care).

**TABLE S.3** Parameters for Accomplishing Vital and Urgent Ocean Drilling Science Research Priorities

| | **Deep Water** >3,000 m of water | **Deep Penetration** >30 m into sediment/rock | **Continuous Records from Cores** No unknown gaps in recovery | **Ephemeral Properties** Porewater, Magnetics | **Borehole Observatory/ Instrumentation** Chemistry, Physics, Biology, Geology | **Logging** Downhole tools after coring | **Ice Strengthened (not icebreaker)** |
|---|---|---|---|---|---|---|---|
| **Ground Truthing Climate Change** | **R**<br>Records are from all world's ocean environments | **R**<br>Required for old records AND younger records with high resolution (high sed rates) | **R**<br>Requires multiple recoveries per location with intentional offsets | **R**<br>To document potential alteration of physically recovered material | **NN** | **G/NN**<br>Cannot replace continuous records from cores | **R**<br>Dependent on target of interest |
| **Evaluating Marine Ecosystem Responses to Climate and Ocean Change** | **R**<br>Records are from all world's ocean environments | **R**<br>Required for deep-time biotic events | **G**<br>Required to infer timing and tempo of ecological response to environmental perturbations | **G**<br>Chemical fluxes upward from the seafloor to the ocean are indicators of and sustain deep life | **G/NN** | **NN** | **R**<br>Dependent on target of interest |
| **Monitoring and Assessing Geohazards** | **R**<br>Continental margin and trenches, deep-water records of volcanic ash, midocean ridge relationships | **R**<br>Deep seismogenic zones, old records of recurrence | **R/I**<br>Dependent on target of interest (yes to temporal earthquake records, eruption records), multiple recovery not required | **R/I**<br>Dependent on target of interest | **R**<br>Very strong requirement for time-dependent hazards assessment | **I/R**<br>Perhaps could replace continuous core recovery in certain cases | **G**<br>Dependent on target of interest |
| **Exploring the Subseafloor Biosphere** | **R**<br>Organic matter supply varies with depth and distance from shore | **R**<br>Habitability in low-energy substrates | **G/NN**<br>Depth and age (only) necessary | **R**<br>Very strong requirement | **I**<br>Dependent on target of interest | **NN** | **G/NN**<br>Dependent on target of interest |
| **Characterizing llk Tectonic Evolution of Ocean Basins** | **R**<br>Midocean ridges and old oceanic crust are deep water | **R**<br>Establishing crustal boundaries and accessing crust beneath buried sediments | **R/I**<br>Depending on target of interest (see Logging) | **G/NN** | **R**<br>Very strong requirement | **R**<br>Due to challenging recovery of hard rock | **I/G** |

NOTES: G = good if a byproduct of a primary driver; I = important, but not required by itself; NN = not necessary; R = required.

### Workforce Needs

A diverse, equitable, and inclusive workforce in ocean science and engineering is an infrastructure component fundamental to the advancement and future success of scientific ocean drilling. With that underpinning, a trained workforce skilled in planning, collection, analysis, and archiving of scientific samples and stewardship of data has been and will continue to be critical to the future of all ocean sciences, and ocean drilling contributes significantly to this goal. Some of the highly specialized positions in the drilling program will likely be lost with the closure, or even temporary cessation, of the U.S. scientific drilling program.

### Managing the Future Scientific Ocean Drilling Enterprise

A nimble and focused management structure is key to a sustainable and successful future U.S.-based scientific ocean drilling effort. Management and staffing requirements will depend on the nature of the U.S. program design—whether, for example, a program moving forward is one focused on using contracted MSPs or an acquired (through long-term lease or build) globally ranging dedicated vessel. Potential questions to consider in determining the scale and scope of a future management structure include:

- What are the short- and long-term financial, scientific, operational, and leadership advantages and disadvantages to various platform options?
- Which measurements are required to be made on the platform?
- How many shore- and ocean-based scientific and technical staff are needed and what is the most appropriate staffing model to meet the needs of the program?
- What is the minimum advisory structure needed for the program?

Given the resource constraints affecting scientific ocean drilling, the answers to these questions could help trim the cost of operations and maintenance, which have plagued the ocean sciences community not only for scientific drilling, but also for other critical infrastructure. Identifying the minimum required program capabilities to advance vital and urgent scientific goals would help to facilitate a stable, successful, dynamic, and sustainable U.S. scientific ocean drilling research program.

## CONCLUSION

The rapid pace of climate change and related extreme events, sea level rise, changes in ocean currents and chemistry threatening ocean ecosystems, and devastating natural hazards such as earthquakes are among the greatest challenges facing society. By coring the past to inform the future, U.S.-based scientific ocean drilling research continues to have unique and essential roles in addressing these vital and urgent challenges.

# 1

# Introduction

This chapter sets the stage for the report by first discussing the importance of scientific ocean drilling in understanding Earth system science and how Earth has evolved over time and behaved during past climate perturbations. The chapter describes the current state of the U.S. scientific ocean drilling program including recent science planning activities to chart out the path forward for the program which provides a segue into the context and rationale behind conducting this study. The chapter ends with a description of the study process and then walks the reader through the report structure.

## THE IMPORTANCE OF SCIENTIFIC OCEAN DRILLING

Astronomers use space telescopes to investigate deep space and look back in time to better understand the nature of galaxies and how they change. Similarly, ocean scientists use vessels equipped with specialized drilling and sampling devices to collect sediment, rocks, and other materials and to leave behind observatories hundreds of meters beneath the seafloor. The data collected using these materials and observatories reveal how Earth, its climate, environment, and life are linked through interconnected processes that operate over a range of temporal and spatial scales. Scientific ocean drilling has enabled critical data, analysis, and insight on how Earth's oceans, atmosphere, and climate have evolved from hundreds of millions of years ago to the last century, and on regional to global spatial scales. It has led to the innovation of technological tools and analytical methods and the cultivation of a highly skilled scientific and engineering workforce that addresses societally relevant Earth system challenges through research, exploration, graduate and undergraduate education, and training. Over its 56-year history, scientific ocean drilling has brought together a global community of scientists and scholars toward these goals. In so doing, the program has facilitated partnerships among the United States and other nations and established U.S. leadership in ocean exploration and research.

Cores drilled from the subseafloor offer a record that extends into Earth's history three orders of magnitude deeper than do ice cores. Undisturbed by erosion or human development, subseafloor records are gleaned from carefully selected locations, as deep as 1,800 m below the seafloor. These cores are unique and highly valued archives of Earth's history. The chemical, physical, biological, and geologic properties of collected samples have improved understanding of key episodes and tipping points in Earth's history, such as abrupt warming events—including rates of onset and recovery—and causes of and connections to perturbations in the global carbon cycle, ocean circulation, and sea level, as well as the collapse of ice sheets. These subseafloor samples also reveal the past health and habitability of the ocean, providing insight into periods of widespread ocean acidification, deoxygenation, and fluctuations in nutrient availability, which affect the stability and composition of marine ecosystems. Data on these recorded changes, preserved in subseafloor sediment, are input into predictive Earth models, helping researchers understand how the climate and ocean could change in the future.

The subseafloor sedimentary layers and basement rock also provide other unique and valuable information. Instrumented monitoring systems, installed in subseafloor drilling locations where tectonic plates meet, offer valuable opportunities to understand time-dependent natural geohazards, such as earthquakes, tsunamis, and volcanic eruptions. Such monitoring systems can provide unique and timely data (if conveyed in real time) on the mechanisms that control the recurrence, timing, and possible precursors of these geohazards.

Moreover, the subseafloor sediment and ocean crust are reservoirs of energy and matter; they are also home to a complex, active, globally spanning ecosystem of microbial life that is only beginning to be explored and understood. Studying the limits and adaptations of life in this deep biosphere may be the best opportunity to understand the possibility of life on other planets, as well as to unlock the potential discovery of novel biological molecules for use in biotechnology and biomedical applications.

These areas of research rely on a well-established scientific ocean drilling domain. Led by the United States, with a history that extends back to 1958, scientific ocean drilling has evolved through different program phases and international collaborations (Figure 1.1). Partnerships forged through scientific ocean drilling activities serve as a model for interpersonal and multinational cooperation. The subseafloor data and materials recovered from past scientific ocean drilling expeditions provide opportunities for scientists around the world to obtain samples and ask new and exciting questions, and for educators to train and mentor the next generation of the science, technology, engineering, and mathematics (STEM) workforce, in order to address existing challenges that lie at the intersection of the Earth's systems and human society (Box 1.1).

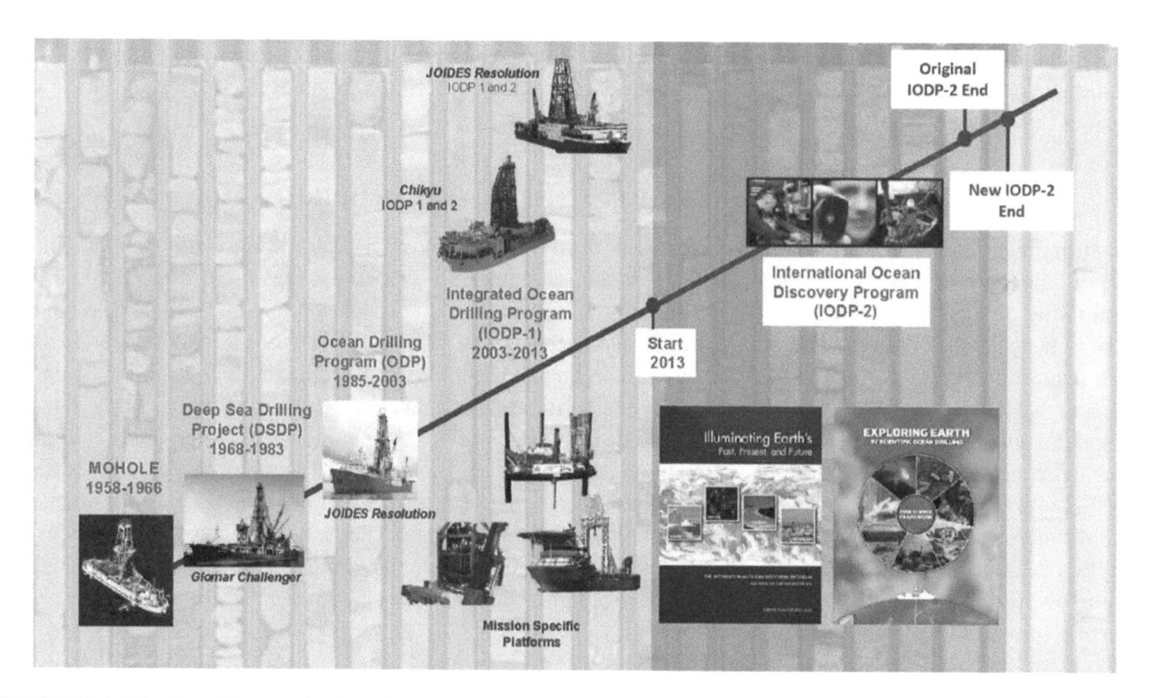


**FIGURE 1.1** The long history of scientific ocean drilling. NOTES: The United States and its flagship vessel, the *JOIDES Resolution*, have been the primary drilling platform from the beginning of the Ocean Drilling Program phase. During the Integrated Ocean Drilling Program (IODP-1) and the International Ocean Discovery Program (IODP-2) phases, capabilities expanded by adding international partners operating additional platforms. The current phase of the scientific ocean drilling program is shaded in blue. Cover images of the science plan for the current program phase and a plan guiding future scientific ocean drilling are included.
SOURCE: Modified from Gilbert Camoin, ECORD Managing Agency.

## BOX 1.1
## Connecting Scientific Ocean Drilling to Other Fields

The research enabled by scientific ocean drilling is highly cross-disciplinary and, by design, is often intended to address questions in a wide range of disciplines. Detailed reconstructions of variations in the ocean and climate have contributed significantly to understanding of Earth system dynamics, particularly those systems that have relatively long response times to natural changes. The following are three examples of connections between scientific ocean drilling and other fields.

**Physical Oceanography:** With the development of advanced ocean circulation models capable of simulating the meridional overturning circulation (MOC) (see Figure 1.2), a system of ocean currents carrying water around the globe, it was discovered that with increased freshwater input at high latitudes (such as from the melting of glaciers), the circulation would abruptly transition from one stable state to another (Rahmstorf, 2002; Weijer et al., 2019). The threshold conditions for these transitions became known as *tipping points*, which were then defined in parallel reconstructions of deep-sea chemical distributions for the last glacial maximum. The paleoceanography reconstructions, based in part on sediment cores recovered by the International Ocean Discovery Program, showed that the MOC operated differently before abruptly transitioning to the present-day circulation pattern (e.g., Böhm et al., 2015; Curry and Oppo, 2005; Roberts et al., 2010). This discovery was particularly timely, as these and similar ocean circulation models, when coupled to atmospheric models and forced by greenhouse gas emission scenarios, have projected a possible future collapse of the MOC with significant impacts on regional climate.

**Ocean Chemistry:** The changes in chemical distributions in the ocean, as reconstructed from sediment cores, contributed to efforts to understand global carbon cycle dynamics, the role of biological diversity, and functional niches in biogeochemical cycles. For example, the observations of the glacial–interglacial changes in ocean chemistry, along with other case studies, have proved critical to elucidating the role of the biological pump (the process by which biologically produced carbon-rich particles sink deep into the ocean interior and sediments) in amplifying the rise and fall of atmospheric $CO_2$ over glacial–interglacial cycles, or the role of carbonate buffering in dampening changes in pH, where ocean acidification, rock weathering, and seafloor carbonate deposition are permanent sinks for anthropogenic $CO_2$.

**Astrobiology:** A growing understanding of the deep-subseafloor biosphere and the conditions under which such life exists is contributing to the field of astrobiology, particularly as it pertains to the search for habitable planets. Such planets could host conditions similar to the deep biosphere, allowing for both biotic and abiotic organic synthesis. These planets would likely have oceans and active tectonics with the energy to drive life-sustaining chemical reactions within or near the seafloor. The exploration of life in Earth's subseafloor biosphere permits experimentation with the basic constituents that enable living things to emerge. This also allows the development of instruments and apertures necessary to explore and discover life in planets with habitable zones. Synergistic activities between planetary science researchers and scientific ocean drilling researchers aim to explore such possibilities.[a] This collaborative science aligns with a memorandum of understanding between the National Aeronautics and Space Administration and the National Science Foundation on the achievement of mutual research activities to advance space, Earth, and biological sciences.[b]

[a] See https://www.hou.usra.edu/meetings/oceandrilling2024.
[b] See https://www.nasa.gov/wp-content/uploads/2021/01/2020_nasa-nsf_mou.pdf?emrc=b5475f

*continued*

**BOX 1.1 Continued**

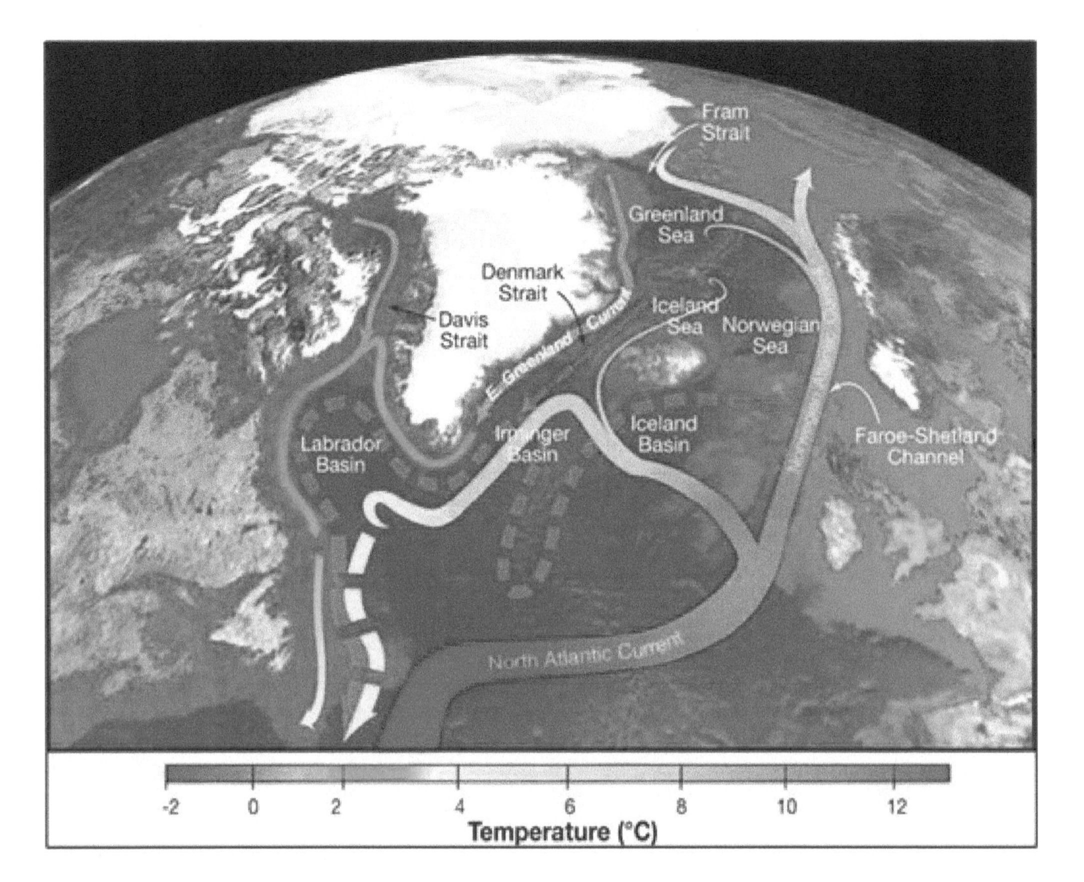


**FIGURE 1.2** Schematic circulation of surface currents (solid curves) and deep currents (dashed curves) that form portions of the Atlantic meridional overturing circulation (AMOC). NOTES: Analysis of core samples and data from scientific ocean drilling expeditions has informed models of AMOC operations during different climate states.
SOURCE: William Curry, Woods Hole Oceanographic Institution/Science/USGCRP, permission for use via Creative Commons 3.0.

## A CRITICAL JUNCTURE

The need to obtain vital records from the subseafloor in carefully selected locations is ongoing and urgent. Scientific ocean drilling is now at a time of consequential transition, which is a result of a confluence of three main factors:

1. The current phase of the drilling program, the International Ocean Discovery Program (IODP-2), ends in 2024.
2. The only dedicated (long-term lease), versatile globally ranging vessel for IODP-2, the aging (45-year-old) research vessel *JOIDES Resolution* will be out of compliance with modern maritime environmental

**TABLE 1.1** Expedition Statistics for the International Ocean Discovery Program (2013–2024)

| | *JOIDES Resolution* (United States) | Mission-Specific Platforms (European Consortium) | *Chikyu* (Japan) | Program Total or Record |
|---|---|---|---|---|
| Expeditions Completed | 46 (81%) | 6 (10%) | 5 (9%) | 57 |
| Sites Visited | 193 (85%) | 28 (12%) | 6 (3%) | 227 |
| Holes Drilled | 530 (84%) | 80 (13%) | 18 (3%) | 628 |
| Deepest Hole Penetrated[a] (m below seafloor) | 1,807 | 1,335 | 1,180 | 1,807 (max) |
| Shallowest Water Depth (m) | 87 | 20 | 1,939 | 20 (min) |
| Deepest Water Depth (m) | 5,012 | 8,023 | 4,776 | 8,023 (max) |
| Core Recovery (m) | 78,067 (95%) | 3,373 (4%) | 1,085 (1%) | 82,525 |

[a] The deepest hole drilled for all of scientific ocean drilling history was 3,262.5 m below seafloor at Site C0002 using the *Chikyu*. Coring at this site occurred during multiple expeditions across two phases of the program; most of the coring occurred during Expeditions 326, 338, and 348 (reaching a depth of 3,058.5 m). The hole depth was extended during Expedition 358 (Tobin et al., 2020).
NOTES: Data through Expedition 399, June 2023. Percentages of total per category provided where appropriate.
SOURCE: Updated by Mitch Malone, modified from IODP, n.d. *JOIDES* Resolution photograph by Bill Crawford.

and safety requirements in 2028. *JOIDES Resolution*[1] is the "workhorse" of international scientific ocean drilling (Table 1.1) and is operated by the United States.

3. The contributed funding from international partners for the operation of *JOIDES Resolution* has decreased over time, making a sustainable budget model challenging. This is a chronic problem, as noted in a previous report by the National Academies of Sciences, Engineering, and Medicine, *Sea Change: 2015-2025 Decadal Survey of Ocean Sciences* (NRC, 2015), and is unlikely to be reversed.

These factors contributed to the decision by the National Science Foundation (NSF, 2023)[2] to end the cooperative agreement with Texas A&M University for *JOIDES Resolution* operations. While no scientific drilling with *JOIDES Resolution* will be supported, NSF has pledged to support some aspects of drilling-related science until 2028, such as the Gulf Coast Repository and data storage, and the necessary management of postexpedition science activities (e.g., publications of expedition reports) through the last IODP-2 expedition (scheduled for 2024).

In addition, the international landscape for scientific ocean drilling is changing rapidly. The European Consortium for Ocean Research Drilling and Japan—which are the primary operational partners with the United States for the current phase of the scientific ocean drilling program—have released a joint vision for future scientific ocean drilling, known as IODP[3] (Camoin and Eguchi, 2023; Kinkel, 2023). The new vision would be a partnership between these ECORD and Japan and potentially other countries currently in IODP-2, and would include

[1] "JOIDES" is an acronym for the Joint Oceanographic Institutions for Deep Earth Sampling (see https://joidesresolution.org/about-the-jr).
[2] See https://www.nsf.gov/news/news_summ.jsp?cntn_id=306986&org=OCE.

the combination of European-led ocean drilling operations of mission-specific platforms, as during IODP-1 and IODP-2, which contracts vessels and other platforms on an expedition-by-expedition basis, and the operation of the very large, specialized, Japanese drilling vessel, *Chikyu*, specifically designed for drilling very deep into ocean sediments and rocks in and near Japanese waters. Currently, the United States has not joined IODP³, nor is there a dedicated globally ranging drilling vessel that can take on the activities that will be lost with the decommissioning of the *JOIDES Resolution*.

A further change to the international landscape is that China is independently developing a plan for its future leadership in scientific ocean drilling, centered around a new drilling vessel and using existing smaller platforms. The new Chinese flagship deep-sea drilling vessel, the *Mengxiang*, was launched for sea trials in December 2023. China plans to make berths available to the developing nations that are part of their Belt and Road initiative (Yand and Shumei, 2024).

Given these national and international factors, **scientific ocean drilling is at a critical juncture, and the future of U.S. operational and scientific leadership and participation in scientific ocean drilling research is at risk** (Figure 1.3).

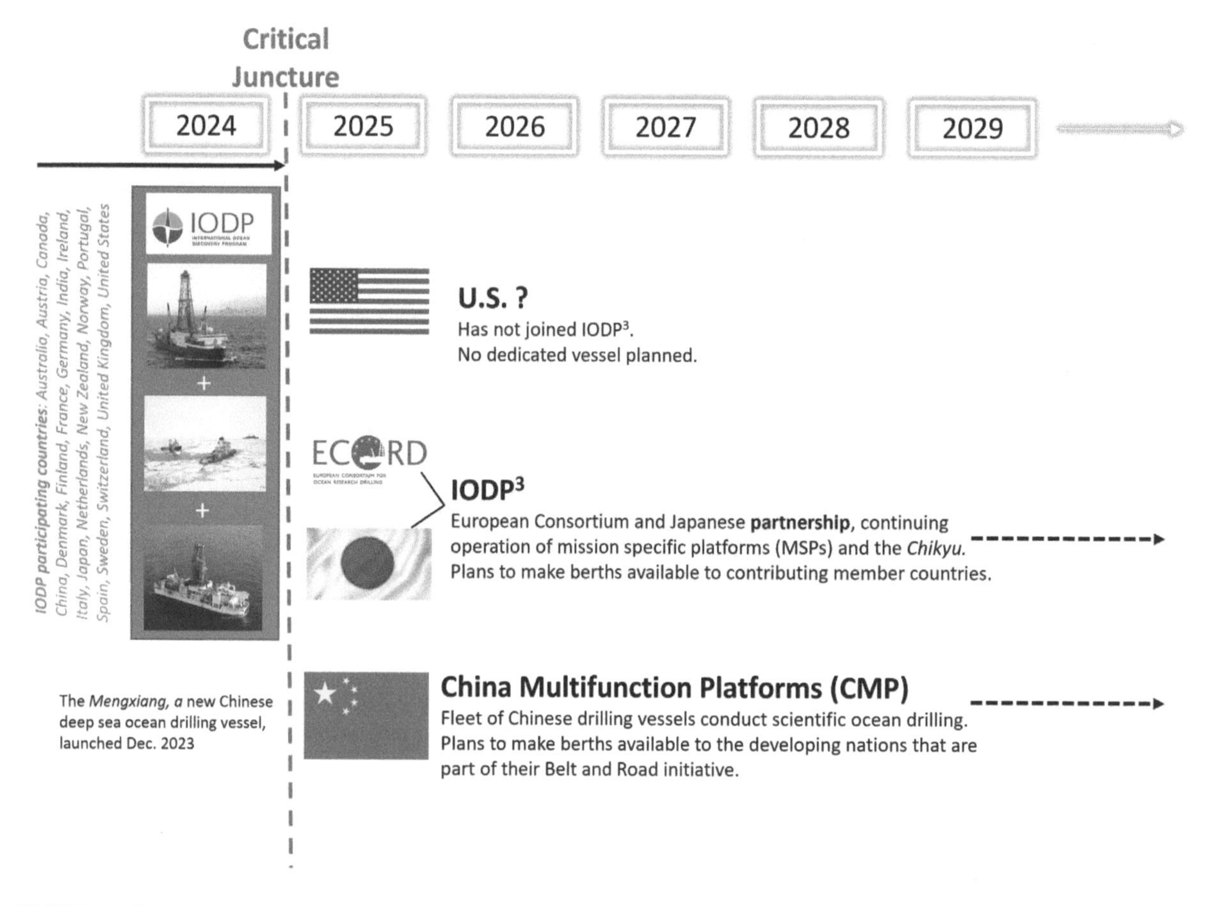


**FIGURE 1.3** A critical juncture in the future of U.S. and international scientific ocean drilling. NOTES: ECORD = European Consortium for Ocean Research Drilling; IODP = International Ocean Discovery Program
SOURCES: Camoin and Eguchi, 2022, slides 4–5; ECORD, 2023; IODP, n.d.b.; Yand and Shumei, 2024.

## PREPARING FOR THE NEXT STAGE OF SCIENTIFIC OCEAN DRILLING

With the current phase of the program scheduled to conclude in 2024, the U.S. and international communities have been planning for the future of scientific ocean drilling. Efforts began in 2018 with a series of workshops to conceptualize a new scientific ocean drilling program, with emphasis on identifying priority science areas, strategic infrastructure, and collaborations. Approximately 650 scientists participated in six planning workshops held in the United States, Japan, India, Europe, Australia and New Zealand, and China.[3] The convergence of the community-wide perspectives from these planning workshops ultimately led to the development of the *2050 Science Framework: Exploring Earth by Scientific Ocean Drilling* (Koppers and Coggon, 2020), referred to hereafter as the *2050 Framework*.

### 2050 Science Framework: Exploring Earth by Scientific Ocean Drilling

The purpose of the *2050 Framework* is to guide multidisciplinary subseafloor research on the interconnected processes that characterize the complex Earth system and shape the planet's future. With a 25-year outlook, it serves as a community-developed science plan subsequent to the *2013–2023 IODP Science Plan*, which was extended to 2024 (IODP, 2011).

The *2050 Framework* identifies a series of strategic objectives as well as a set of long-term initiatives. The seven strategic objectives focus on understanding the interconnections within the Earth system (Figure 1.4). The initiatives (referred to in the report as flagship initiatives) are multidisciplinary research endeavors that combine goals from multiple strategic objectives; these are summarized in Box 1.2. Additionally, four "enabling elements" were defined for enhancing scientific output and impact. These include education and outreach; collaborative research with continental drilling programs (e.g., International Continental Scientific Drilling Program, U.S. Continental Scientific Drilling Facility), collaborative research with space agencies (e.g., National Aeronautics and Space Administration); and advancing technological developments and big data analytics.

### Community-Identified Science Mission Requirements

In 2022, following the publication of the *2050 Science Framework*, and recognizing that the *JOIDES Resolution* was approaching the end of its operational utility, NSF requested that the U.S. science community provide further input on U.S. science priorities and regional foci for subseafloor drilling and define necessary vessel design requirements to meet these priorities. In response, U.S. community input was gathered via online surveys and a series of workshops, resulting in the report *Science Mission Requirements for a Globally Ranging, Riserless Drilling Vessel for U.S. Scientific Ocean Drilling* (hereafter referred to as the *SMR* report) (Robinson et al., 2022). The science priorities described in the report include the themes of

- climate change;
- life on Earth;
- natural hazards; and
- the cycling of tectonic plates, energy, and matter.

The geographic areas of interest identified in the *SMR* report (Robinson et al., 2022) span the global ocean, including all ocean basins, with special emphasis on high-latitude settings. Other special settings include continental shelves and slopes, glaciated margins, ocean ridges, and subduction zones and trenches. The *SMR* report concluded that prioritization should go toward working in unexplored locations (geographic and geological record gaps) and recovering cores that are representative of the target geology and microbiome, with an emphasis on characteristically difficult settings (e.g., unconsolidated sediments, glaciomarine sediments, fractured formations, young oceanic crust). The report also identified as a priority obtaining petrophysics logs of drilled holes, especially

[3] Workshop reports of four of the planning workshops can be accessed at https://www.iodp.org/iodp-future/planning-workshop-outcomes.

**FIGURE 1.4** The seven strategic objectives of the *2050 Science Framework* are broad areas of scientific inquiry that focus on understanding the interconnected Earth system.
SOURCE: Ellen Kappel, Geo-Prose, modified from Koppers and Coggon, 2020.

drilling the upper 50–100 m below the seafloor, which is not achievable in the current phase of the program. Technological advancements envisioned in the report include installing more observatories and implementing a greater range of tools for in situ measurements of subseafloor conditions (e.g., fluid flow, slip rates).

The *SMR* report (Robinson et al., 2022) described two categories of vessel design characteristics necessary for meeting the identified science priorities: *foundational requirements*, which are the minimum criteria for a future riserless drilling vessel,[4] and *more robust design features*. The eight foundational requirements are interrelated (Figure 1.5). The report concluded that a vessel—supported by highly trained technical and scientific personnel—that can provide access to sites globally, operate in diverse subseafloor settings, and control specific aspects of the downhole conditions is essential for accessing key geologic and biologic environments, collecting high-quality cores, and establishing subseafloor observatories.

[4] Specific drilling terminology is described in Box 2.1 in Chapter 2.

## BOX 1.2
## *2050 Science Framework* Flagship Initiatives

In 2020, a group of scientists representing 23 countries published a guide on global-scale, interdisciplinary, societally relevant research priorities for the next 30 years of scientific ocean drilling, titled the *2050 Science Framework: Exploring Earth by Scientific Ocean Drilling* (Koppers and Coggon, 2020). From the *2050 Science Framework* derivative[a] pamphlet:

> The 2050 Science Framework guides scientists on important research frontiers that scientific ocean drilling should pursue. It focuses on the many ways in which scientific ocean drilling will increase understanding of the fundamental connections among Earth system components while addressing a range of natural and human-caused environmental challenges facing society.
>
> The Flagship Initiatives comprise long-term, multidisciplinary research endeavors that aim to test scientific paradigms and hypotheses that inform issues of relevance or interest to society. They typically combine research goals from multiple strategic objectives. Their implementation will be shaped by proposals from the scientific community that develop coordinated strategies that include long-term planning, technology development, and innovative applications of existing and new scientific ocean drilling data products.
>
> **Ground Truthing Future Climate Change**. By collecting the robust data required for reconstructing global climate evolution over extended geologic time periods, scientific ocean drilling will provide information that is critical for improving climate model performance.
>
> **Probing the Deep Earth**. By penetrating deep within oceanic crust, scientific ocean drilling will lead to a better understanding of Earth's formation and evolution and the connections between tectonics, earthquake and volcanic hazards, climate, and the planet's habitability.
>
> **Assessing Earthquake and Tsunami Hazards**. By acquiring samples and deploying instruments in offshore and nearshore fault zones, scientific ocean drilling will enable more reliable assessments of the risks posed by major earthquakes and tsunamis and will facilitate improved hazard preparedness and response.
>
> **Diagnosing Ocean Health**. By retrieving sedimentary records that preserve key information about past responses of biological activity to natural cycles and catastrophic events, scientific ocean drilling will enable a more informed assessment of the expected rates, duration, and magnitudes of future ocean health deterioration.
>
> **Exploring Life and Its Origins**. Scientific ocean drilling and monitoring in borehole observatories will advance research into the distribution and limits of deep microbial life, novel microbes and their biotechnological applications, the emergence and evolution of life on Earth, and the possibility of life on other worlds.

[a] See https://www.iodp.org/docs/iodp-future/1087-2050-science-framework-pamphlet/file.

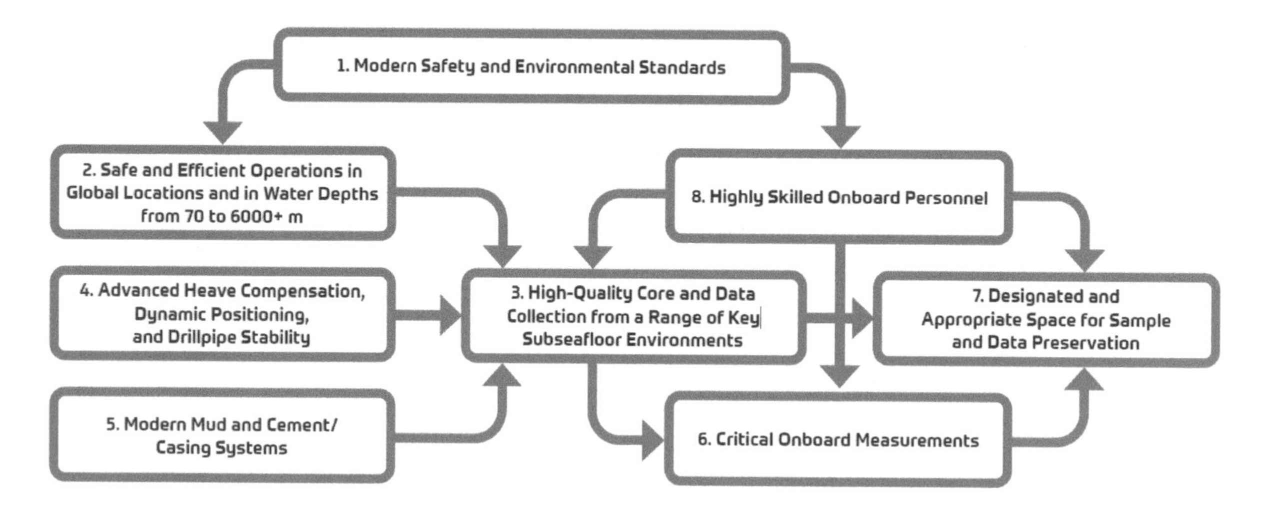


**FIGURE 1.5** Dependencies and relationships of the foundational science mission vessel requirements.
SOURCE: Robinson et al., 2022.

### Workshops: Envisioning Science Communication and Outreach

In addition to science and infrastructure planning meetings and reports, the U.S. scientific ocean drilling community also held a series of workshops in 2022 designed to inform science communication and outreach for future scientific ocean drilling. The workshops focused on three areas: engaging the public (White et al., 2021), informing policy makers (Cotterill et al., 2021b), and preparing the next generation of scientists (Cotterill et al., 2021a). Key premises of these workshops were that (1) robust science communication, outreach, and education are important enabling elements for the future of U.S. scientific ocean drilling, and that (2) consideration of diversity, equity, and inclusion provides an important lens through which to consider these topics. Findings were wide ranging, but included the recognition that there is a need to build greater awareness of the important contributions of scientific ocean drilling to issues that affect science policy and society, and that there is great value in utilizing approaches that capture public imagination and interest and provide new understanding of Earth. Findings also identified the value of utilizing core samples and data in course-based research experiences for undergraduate students and in short courses and workshops for graduate students and early-career researchers; such opportunities can help develop critical scientific and transdisciplinary skills that contribute to building the STEM workforce.

## STUDY ORIGIN AND PURPOSE

In addition to efforts underway by the U.S. and international scientific ocean drilling communities to prepare for the demobilization of the *JOIDES Resolution*, the conclusion of the current drilling program, and the transition to a new phase of ocean drilling, NSF also asked for input from the broader U.S. ocean science community. In response, the National Academies formed the Committee on the 2025–2035 Decadal Survey of Ocean Sciences for the National Science Foundation. The goal of the present study, an interim report for the Decadal Survey of Ocean Sciences (DSOS), is to provide a broader perspective on high-priority science needs that require ocean drilling, and the resources and infrastructure needed to accomplish such important research. The committee's task for the interim report is presented in Box 1.3 (the full DSOS statement of task is included in Appendix A). In consultation with NSF, the committee has chosen to address the statement of task by highlighting high-priority science themes (i.e., areas, topics), rather than specific scientific questions. Because scientific ocean drilling is at a time-sensitive juncture, NSF requested that the committee produce an interim report focused solely on scientific ocean drilling

**BOX 1.3**
**Statement of Task**

The Committee will produce an interim report to provide advice to NSF OCE on the resources and infrastructure available to address high priority research questions requiring scientific ocean drilling. The interim report will cover the following:

1. Based on previous reports, assess progress on addressing high priority science questions that require scientific ocean drilling and identify new, if any, equally compelling science questions that would also require scientific ocean drilling.
2. Of the unanswered scientific questions, which could be addressed through the use of existing scientific drilling assets including sediment or rock core archives and existing platforms, and which questions would require new infrastructure or sampling investments?

ahead of the DSOS final report (Appendix A). This report therefore aims to fulfill the two tasks listed in Box 1.3 and does not include recommendations to NSF, which fall outside the scope of this report. Over the next year, as the committee continues work on the final DSOS report, the findings and conclusions from this study will inform the committee's recommendations on the broader ocean science research and infrastructure portfolio for NSF.

## STUDY APPROACH

The committee's deliberations and resulting conclusions were informed by its collective expertise; review of scientific literature; and several mechanisms for hearing directly from the broad scientific ocean drilling community, including public meetings and a virtual town hall. The main public information gathering (see Appendix B for agenda and participant list) consisted of a 2-day meeting on scientific ocean drilling program research priorities and required infrastructure. Thirty guests at all career stages—representing three subcategories of scientific ocean drilling: solid Earth dynamics, climate and environment, and health and habitability—were invited to participate in person. Virtual participation was open to anyone who registered, and roughly an additional 150 individuals participated online.

The committee also opened a virtual town hall for the community to submit their thoughts on ocean science research priorities broadly, as well as those relevant to scientific ocean drilling. The virtual town hall remained open for 2 months and collected input from 94 participants representing voices from early-, mid-, and late-career scientists and practitioners. Roughly 50 percent of respondents were in the late-career stage, and the remaining were split evenly between the early- and mid-career stages.

In addition to solicited outreach, the committee received e-mails, letters, and public feedback submitted through the project website, including input specifically from early-career scientists and international scientific ocean drilling programs. All information shared with the committee was considered in report deliberations.

The process for information gathering, deliberations, report writing, and publication occurred over a relatively short period of time. The findings and conclusions from this study are interim; they will be incorporated and referenced in the full report as appropriate, and as a component of the larger ocean sciences priority research portfolio.

## REPORT ORGANIZATION

Chapter 1 provides the context for this study. An overview or "primer" of scientific ocean drilling is provided in Chapter 2, including an overview of how the program works, a history of scientific ocean drilling and a historical summary of the major accomplishments achieved using scientific ocean drilling over the decades.

An assessment of progress made in scientific ocean drilling over the last decade and questions remaining for the future are summarized in Chapter 3. Chapter 4 then examines the priorities in the context of what can and cannot be accomplished with available infrastructure. The committee's main findings are presented throughout the text and conclusions follow the text supporting them.

# 2

# A Primer on Scientific Ocean Drilling

This chapter provides background on the scientific ocean drilling program important for understanding the research priorities and infrastructure needs described for a future program. First, key terms used throughout the report are defined, particularly those important for understanding the difference between platforms and/or mechanisms through which scientific ocean drilling has been conducted over the past decades and may be conducted in the future. This is followed by an overview of the scientific drilling management structure and process for planning and conducting expeditions. The chapter then presents a brief history of ocean drilling, including major accomplishments.

## TERMINOLOGY AND KEY CONCEPTS

For decades, scientific ocean drilling pioneered global-scale interdisciplinary research below the seafloor of the world's ocean. As distinctions between shallow subseafloor coring and deeper drilling-enabled core recovery are important to the capabilities and accomplishments of scientific ocean drilling, definitions and other report-relevant terms are provided in Box 2.1. A geological timeline that spans most of the time periods accessible via scientific ocean drilling is provided (Figure 2.1), along with fundamental data derived from scientific ocean drilling on past global temperatures, sea levels, and global tectonic configurations.

Eight unifying, guiding principles have emerged from the scientific philosophies and practices over the history of the scientific ocean drilling program and has been identified by the U.S. and international scientific ocean drilling program communities (Koppers and Coggon, 2020) as important for any and all drilling programs:

- ***Open access to samples and data.*** Free and open access to samples in core repositories and data in online repositories has been a hallmark of the program. Open access ensures that scientists of all career stages can participate in the program. In practice, samples and data are subject to a 1-year moratorium post-expedition, during which time samples and data are available only to the expedition team, including shore-based participants. After 1 year, all the data are made publicly accessible by the scientists who collected them, and others from around the world can submit requests for core samples to any of the three scientific ocean drilling core repositories (located in Texas, Germany, and Japan) for use in research and teaching and training (IODP, n.d.e).[1]

[1] See https://www.iodp.org/resources/core-repositories. Also, samples from a collaborative sea-to-land coring project are stored in a repository in New Jersey.

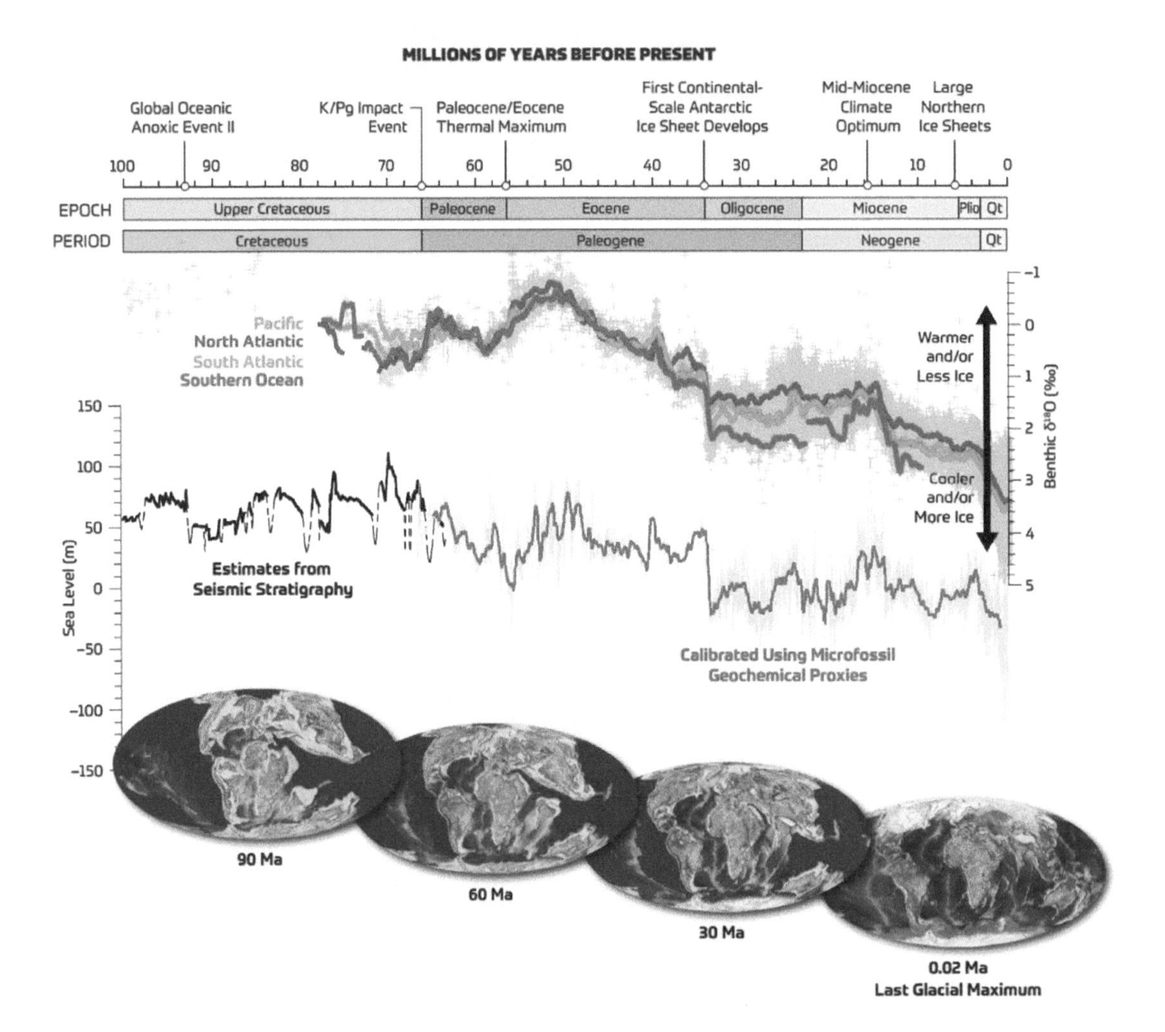


**FIGURE 2.1** Geological timeline of the last 100 million years, as recorded in scientific ocean drilling records from marine sediments. NOTES: Significant paleogeographic reorganization and key climatic events in Earth history accompanied a long-term cooling trend that culminated in the development of permanent ice sheets in the polar regions. The K–Pg (Cretaceous–Paleogene) impact event marks a global mass extinction. The geochemical composition ($\delta^{18}O$) of benthic microfossils shown here provides a tool for reconstructing bottom-water temperatures in different ocean basins and calibrating records of global ice volume and sea level change. The most recent estimates of sea level change for the Cenozoic are shown and are based on a spliced (i.e., combined) record derived from multiple scientific ocean drilling cores, while estimates further back in time are based largely on seismic stratigraphy estimates. Sea level values are relative to a preindustrial baseline (0 m). The global maps show the changing configuration of the continents and ocean through time (as boundary conditions of the climate system), as well as the extent of the cryosphere. Ma = million years ago; Qt = Quaternary; Plio = Pliocene.
SOURCE: Koppers and Coggon, 2020. Illustration by Geo Prose based on Cramer et al., 2009.

## BOX 2.1
## Definitions

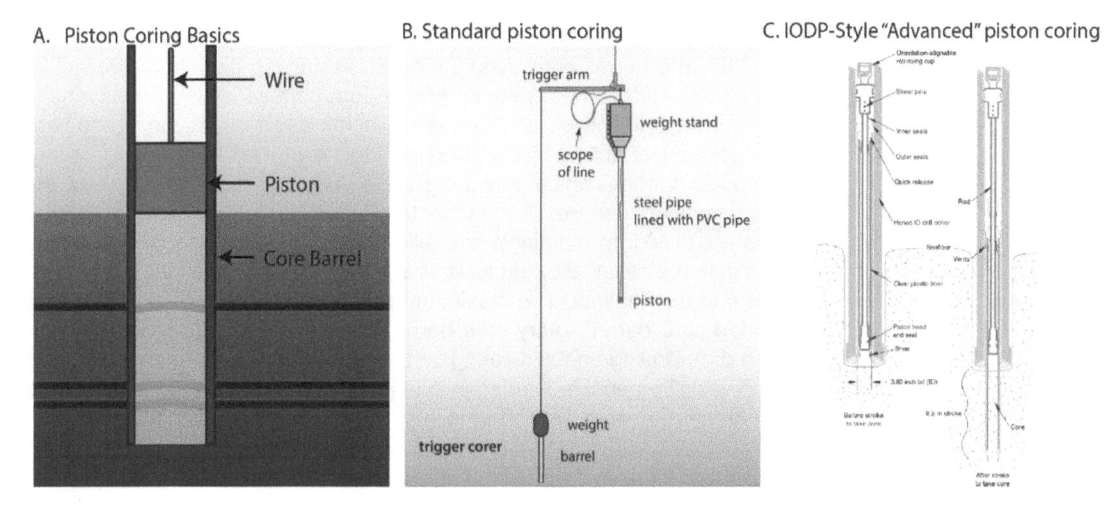


**FIGURE 2.2** (A) The fundamental elements of a piston corer. SOURCE: Maureen Walczak.
(B) Standard piston corer design representing those used in the academic research fleet. SOURCE: David Reinert, CEOAS Communications.
(C) Advanced piston corer design representing those used in scientific ocean drilling.
SOURCE: Baldauf et al., 2002.

**Standard piston corer:** A long, heavy tube with a piston inside that is lowered over the side of a vessel and uses gravity to drive a corer into the seafloor to extract samples of soft (i.e., unlithified) sediment and mud (Figure 2.2A, B). It is unsuitable for recovering hard sediment or rock. The standard piston corer is 3–18 m in length, which is therefore the maximum subseafloor depth for core recovery. In theory, there is no water depth limit for any piston corer.[a] The U.S. Academic Research Fleet (ARF), coordinated by the U.S. University-National Oceanographic Laboratory System (UNOLS), uses this type of coring tool on board its vessels. The *JOIDES Resolution* and other International Ocean Discovery Program platforms are not configured to use this corer.

**Giant piston corers:** Longer and heavier than standard piston corers; also used to core soft sediment and mud and unsuitable for recovering hard sediment or rock. The giant piston corer is 18–60 m in length, with the maximum length only available through a commercial purveyor (e.g., Ocean Scientific International Ltd [OSIL]). Currently, no vessels in the ARF can handle systems of this size.[b] The maximum subseafloor depth for giant piston coring with the current ARF is 40 m; however this limit is approachable only in a few environments based on UNOLS Safety Committee guidelines and cores up to 30-m subseafloor depth is more the norm. A long core system designed for deployment on R/V *Knorr* in the early 2000s had an effective maximum length of 50 m, but this vessel (and coring system) is no longer available in the ARF. Giant piston corers are also known as jumbo piston corers.

**Drilling-enabled advanced piston corers:** Part of a combined drilling–coring system that enables much deeper (Figure 2.2C) and more continuous core recovery than standard or giant piston corers. This apparatus requires using a drillship, and, given current ARF capabilities, is the only means of coring deeper than 30-m subseafloor. In drilling-enabled coring, sections of pipe are connected, extending from a hole (the moonpool) in the drilling vessel down into the seafloor. At the base of the pipe is a drilling bit (drilling

## BOX 2.1 Continued

tool). An advanced piston corer (APC) core barrel is lowered by a wire inside the drill pipe to the bottom of the pipe. Pump pressure is applied to the drill pipe, which hydraulically advances the inner core barrel 9.5 m into soft sediment and mud. The core barrel, containing the core, is retrieved back to the vessel by the wire. After core retrieval, more pipe is added, the drilling bit and bottom-hole assembly are advanced another 9.5 m, and the process is repeated downward continuously until the sediment becomes too hard (or rock is encountered), at which point other drilling bits and coring tools are used (e.g., rotary coring) (IODP, n.d.a). Significant engineering and technological advances have improved the rate of coring, the fraction of mud recovered per 9.5-m advance (see Box 3.1 in Chapter 3), and the fidelity of the recovered material. Commonly, multiple holes are drilled from a single site, allowing for offset coring to recover the disturbed strata at core breaks, and in some cases allowing for a stratigraphic splice of piston-cored material extending many tens of meters into the seafloor. The maximum depth of drilling-enabled coring from the coring tools (e.g., APC, extended core barrel, rotary core barrel) used in scientific ocean drilling is 1,807-m subseafloor depth (IODP, n.d.d). Drilling-enabled coring can occur at a range of water depths. For example, the maximum water depth of drilling-enabled coring in scientific ocean drilling thus far is 5.7 km for the *JOIDES Resolution*, 8.0 km for mission-specific platforms, and 6.9 km for the *Chikyu* (IODP, n.d.d).

**Seabed lander–based drilling systems:** A robotic drilling rig that is deployed to the seafloor and operated remotely from a research vessel. Seabed corers (also called seabed landers or seafloor drilling rigs) can recover up to 260 m of subseafloor sediment and mud using a multihole operational approach to acquire a single representative sedimentary record. This involves drilling through shallower seafloor strata and shifting core operations to deeper seafloor strata to develop a composite section of 260 m over several holes. The diameter of the recovered core is significantly smaller than drilling from a drillship, and real-time monitoring of core recovery cannot occur. The functional water depth limit for seabed corers is approximately 4 km. An example of a seabed corer is the MeBo200 system developed by the Center for Marine Environmental Sciences (MARUM) at the University of Bremen, Germany (MARUM, n.d.b) (Figure 2.3). Currently there are no seabed corers in the ARF, and using them from the ARF will be possible only with some combination of significant financial investment in the infrastructure and capabilities of the vessels themselves; none could be deployed from any vessel in the fleet as it currently exists.

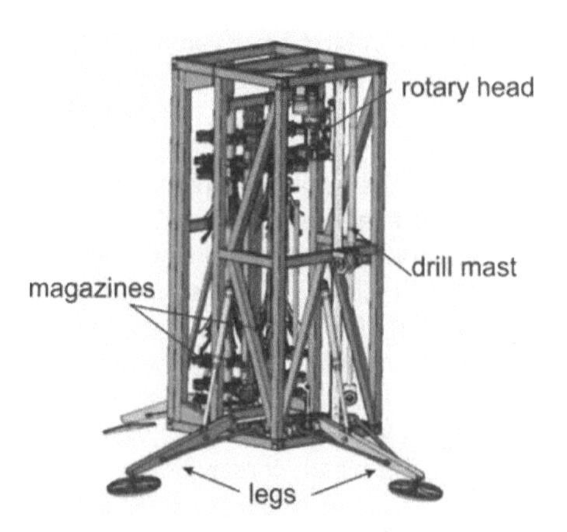


**FIGURE 2.3** Seafloor drill rig MARUM-MeBo200, the second-generation MeBo. NOTE: MeBo = *Meeresboden-Bohrgerät*, German for seafloor drill rig.
SOURCE: BAUER/MARUM, University of Bremen.

*continued*

## BOX 2.1 Continued

**Borehole observatories:** Sensors placed in an open borehole (a drilled hole, such as those drilled to recover subseafloor cores) for long-term monitoring of subseafloor temperature, pressure, strain, tilt, and seismicity, and for collecting borehole fluid samples. A seal is required to isolate the sensors from ocean bottom water; this seal is referred to as a Circulation Obviation Retrofit Kit (CORK) (Figure 2.4). Data collected through these observatories can be transmitted in real time using a subsea cable, or data can be retrieved via remotely operated vehicle from the top of the CORK. Because drilling is required to create the borehole, establishing new CORKs is not possible with the current ARF; they are installed using the *JOIDES Resolution* and other scientific ocean drilling vessels. Borehole observatories have also been installed using the MARUM MeBo seabed coring system for approximately 10 years (Kopf et al., 2015); however, these installations are limited to shallow subseafloor depths (~100 m).

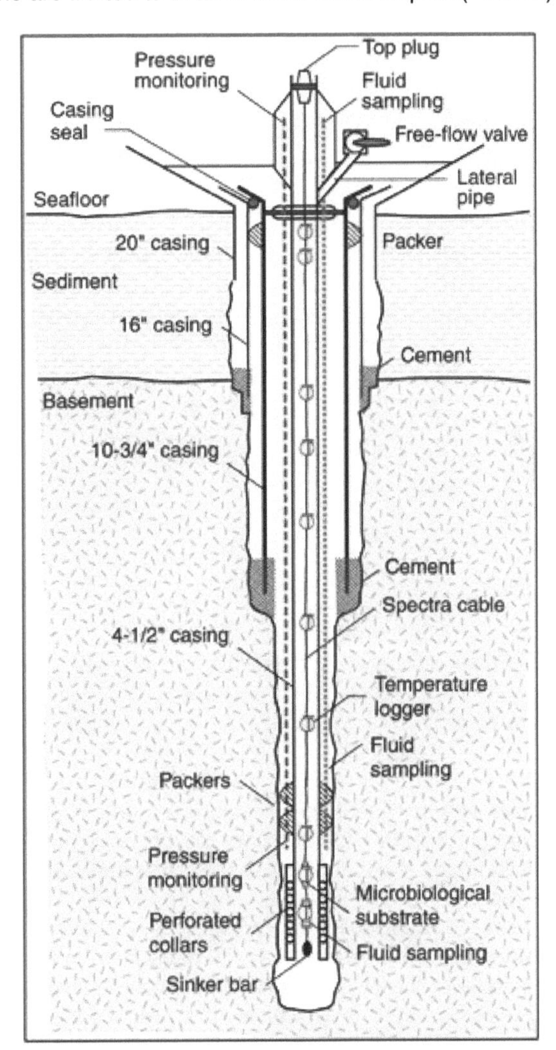


**FIGURE 2.4** Features of new Circulation Obviation Retrofit Kit systems deployed during Expedition 327. SOURCE: U.S. Science Support Program, IODP, Fisher et al., 2011, fig. F2..

*continued*

## BOX 2.1 Continued

**Riserless versus riser drilling:** The *JOIDES Resolution* and its U.S. predecessor, the Glomar *Challenger*, are riserless drillships (Figure 2.5). In short, this means that the hole conditions are stabilized only by drilling mud that is open to the seafloor, and there are no "blowout preventers" or other means of preventing gaseous overpressure. It is the simplest system of drilling yet requires extensive safety planning to avoid hitting hydrocarbon reservoirs or other hazards. The Japanese drilling vessel *Chikyu* deploys riser drilling (Figure 2.5), in which the hole conditions are controlled by drilling mud within a wider pipe outside the drilling pipe, and thus the mud pressure can be controlled, stabilizing the borehole wall. A seafloor blowout preventer controls gas pressure and thus protects the drillship and the environment from gaseous overpressure. Typically, riser drillships can drill deeper than riserless drillships.

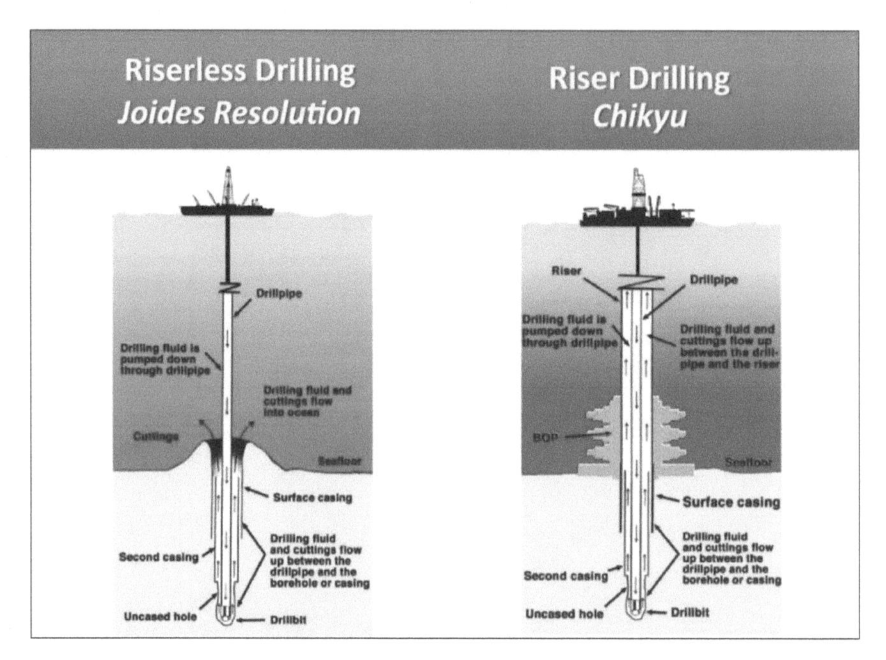


**FIGURE 2.5** Riserless versus riser drilling technology used in scientific ocean drilling. NOTE: BOP = blowout preventer.
SOURCE: Modified from Taira et al., 2014. © JAMSTEC (Japan Agency for Marine-Earth Science and Technology).

**Sedimentation rates:** The rate at which sediment is deposited onto the seafloor, typically in units of centimeters per thousand years (cm/kyr) or kilometers per million years (km/myr). Recovering sediments in an area with low sedimentation rates facilitates recovery of older sediment over a given depth, but with lower age resolution, compared with recovering sediments in an area with high sedimentation, which facilitates higher (i.e., more detailed) age resolution. For example, sediments in the anoxic Cariaco Basin accumulate as fast as 30 cm/kyr; therefore, sediment at 170 m beneath the seafloor is 700,000 years old. In contrast, sediment in the middle of the North Pacific Gyre accumulates at a rate of 1–3 mm/myr, so sediment ~25 m beneath the seafloor is 66 million years old. Paleoceanographers very carefully select the optimal region for addressing past variations in climate in terms of age recovered and resolution of the sedimentary layers. Thick, continuous sequences of sediment layers resulting from high sedimentation

*continued*

**BOX 2.1 Continued**

rates over long time periods are ideal for palaeoceanographic investigations that require detailed sampling and analyses. Because of the relationship between sedimentation rates, subseafloor depth, and age, this often requires collecting sediment cores deep below the seafloor, and benefits from collecting sediment cores from multiple holes to ensure enough material is available for a wide range of analyses.

---

[a] For example, the Oregon State University Marine Rock and Sediment Sampling (MARSSAM) Group cored to 7+ km in a Puerto Rico trench using a standard piston core system from a synthetic line on the research vessel *Neil Armstrong* in early 2022, and the Japanese used a commercial (OSIL) system to core in the Challenger Deep the same year. The use of a synthetic/neutrally buoyant line is essential to avoid wire weight, or to have a sufficiently robust handling system to accommodate the weight of steel. Piston cores are fully mechanical systems, and if used with a trigger arm, there are no depth limitations in theory outside of weight on wire.

[b] Following the 2016 retirement of R/V *Knorr*, the handling systems (wire, winches, cranes, and A-frames) of even the largest vessels in the current ARF are not rated for the working loads associated with 60-m piston coring (e.g., OSIL Giant Piston Corer, see https://osil.com/product/jumbo-giant-piston-corer-18m-60m) and as such, unless existing vessels are modified and/or until new vessels come online, researchers are limited functionally to giant piston coring operations with at most 30- to 40-m recovery.

- ***Standardized measurements.*** The fidelity of scientific research and integrative data analytics depends on high-quality observations and comparable datasets. Standardized measurements and analytical techniques are long-standing practices among the drilling program's technicians and scientists, whether data collection and analysis occur in the floating laboratories onboard the *JOIDES Resolution* or in specialized container laboratories on mission-specific platforms. Measurements and observations are also standardized in the laboratories that host the core repositories. International intercomparability demonstrated by the scientific ocean drilling program has been a model for other ocean science programs.
- ***Bottom-up proposal submissions and peer review.*** Science that is prioritized and ultimately accomplished emerges from visioning, multidisciplinary expertise, and collaboration of scientists from around the world. The bottom-up input occurs at multiple stages and levels in the program. Each phase of the program is guided by an overarching, framing document that describes long-term scientific priorities (e.g., currently this is the *2013–2023 IODP Science Plan*), written and vetted by the scientific community (IODP, 2011). At the expedition level, science objectives result from a proposal process and are subject to multiple stages of peer review for scientific, environmental, and safety considerations. The structure is managed through U.S. and international scientific community panels and boards.
- ***Transparent regional planning.*** For cost and time efficiency, *JOIDES Resolution* expedition scheduling employs a regional planning approach, reducing transits between sequential expeditions (thus maximizing time for conducting research) (IODP, n.d.c). Transparent regional planning by facility boards allows proponent teams to develop proposals in support of strategically timed scientific ocean drilling in a particular area of the global ocean. For example, during the current phase of the International Ocean Discovery Program (IODP-2), the *JOIDES Resolution* maximized time in the Indian Ocean, moved to the western Pacific, then to the South Atlantic, and is now operating in the North Atlantic (Figure 2.6).

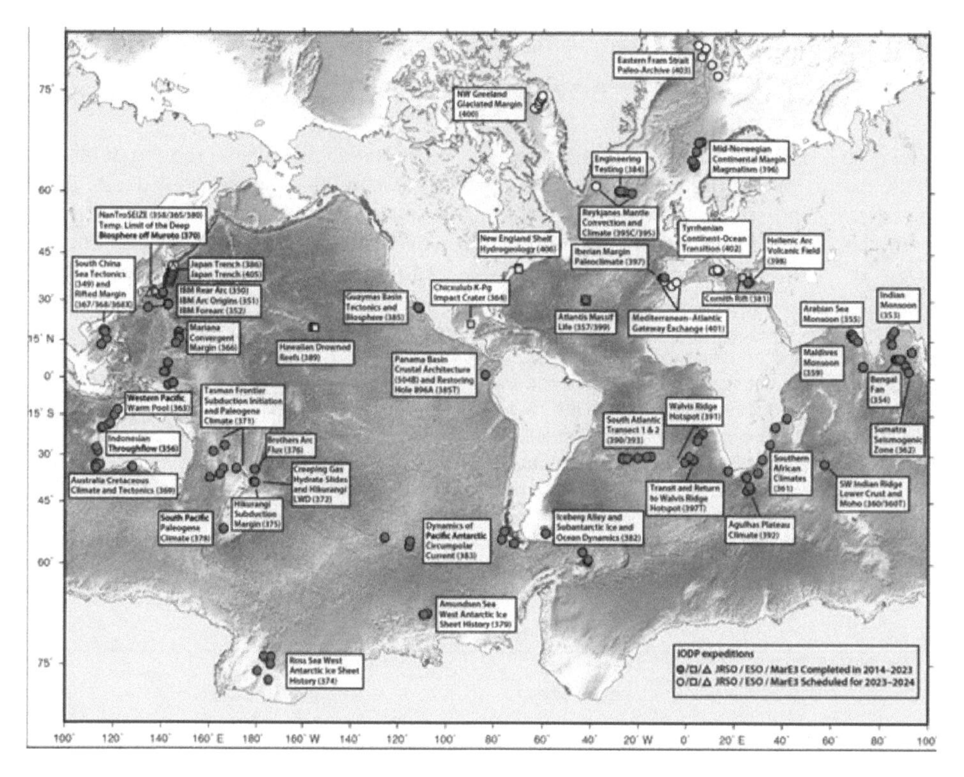


**FIGURE 2.6** Site map of International Ocean Discovery Program expeditions completed in the current phase of the program (2014–2023) and scheduled for the remainder of the program (2023–2024). NOTES: Created July 2023. *JOIDES Resolution* expeditions, managed by the United States (JRSO) are red and yellow circles. Mission-specific platform (MSP) expeditions managed by the European Consortium (ESO) are blue and yellow squares. *Chikyu* expeditions, managed by Japan (MarE3) are green and yellow triangles. Expeditions generally occur in numbered order, although exceptions occur.
SOURCE: *JOIDES Resolution* Science Operator.

- ***Safety and success through location characterization.*** Characterization and evaluation of the drilling location is essential for safety and scientific success prior to the approval of any subseafloor drilling expedition. Site surveys and examination of data collected by other oceanographic research vessels (e.g., the ARF) are fundamental steps of preexpedition planning; these include using seismic reflection and bathymetry data to characterize the seafloor and often gathering shallowly penetrating cores to test penetration. Site surveys ensure that subseafloor drilling data are interpreted in the appropriate scientific context. Careful site selection is critical because the sediments, rock, and other materials in the subseafloor vary from location to location, and achieving the scientific objectives of a given expedition often requires knowledge of site-specific characteristics. For example, seafloor sites with high sedimentation rates and long, continuous (undisturbed) records are essential to long-term, high-resolution analyses of past climate, ocean circulation, and ecosystems changes, and are therefore often targeted when drilling for cores and collecting samples.
- ***Regular assessments.*** Assessment of scientific progress and operational implementation is fundamental for any scientific program to adjust and advance as the goals of science and needs of society evolve. In scientific ocean drilling, regular operational assessments occur at multiple junctures and scales: the National Science Foundation (NSF) conducts annual site visits to the *JOIDES Resolution* Science Operator facility and receives annual evaluation reports completed by co-chief scientists (IODP, n.d.f). Additionally, post-expedition evaluations are completed on the scientific operations by science party members.

- ***International collaboration.*** The scientific ocean drilling community is internationally integrated (Figure 2.7), featuring 21 member countries in the current phase of the scientific ocean drilling program and collaborators in many other countries. Expeditions are always made up of international science teams, with the number of berths allotted dependent on each country's financial contributions to the program. Scientists are often eager to continue the international collaborations that took place onboard once onshore and when analyzing samples.
- ***Workforce diversity and inclusion.*** The scientific ocean drilling community values a diverse and inclusive scientific and technological workforce. With time, the program and the workforce culture has evolved to be more inclusive (Box 2.2). At present, **there is typically equity in the number of women and men in expedition science teams. On average, early-career scientists and graduate students comprise two-thirds of expedition science team members and are mentored by senior scientists**. Cultural and multinational diversity exists throughout, as a direct result of the scientific ocean drilling program (and each expedition) being structured as an international program. Removing barriers and promoting opportunities to further broaden participation is an ongoing effort and varies by country.

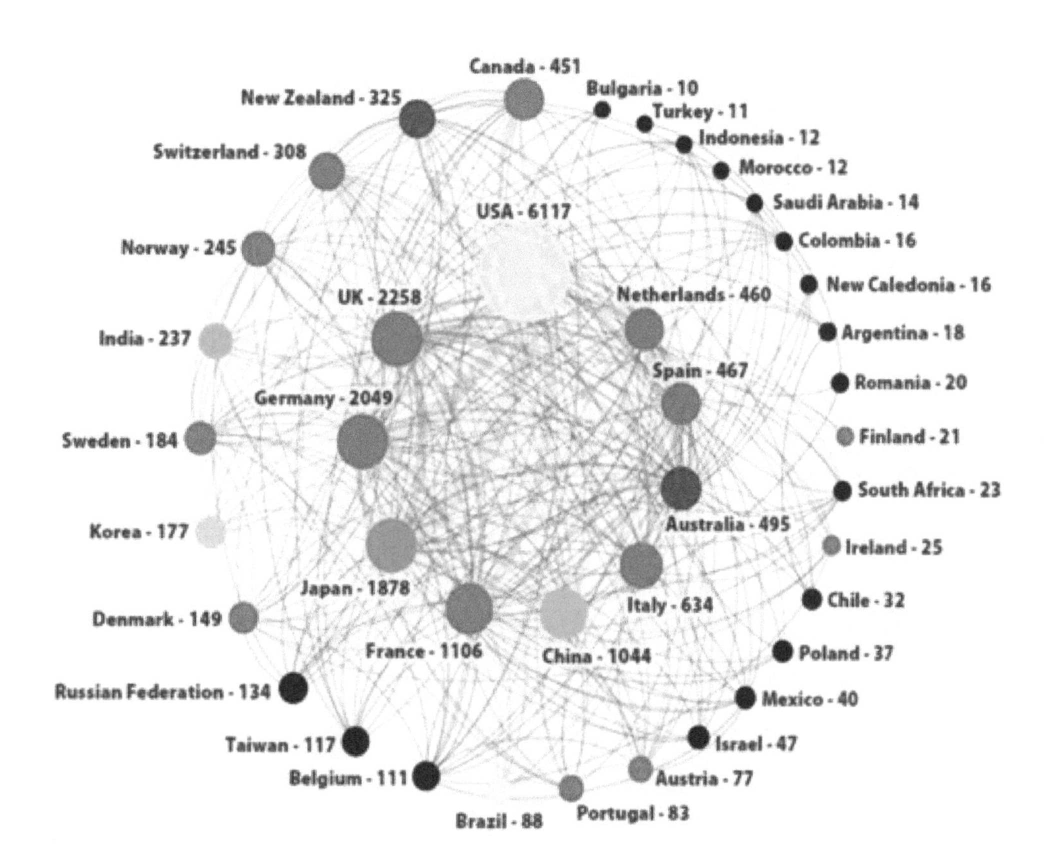


**FIGURE 2.7** Coauthored networks of scientific ocean drilling–related peer-reviewed journal articles from 2003 to 2021.
SOURCE: U.S. Science Support Program, 2022.

## BOX 2.2
## Science Supported by a Diverse Workforce

During the current phase of scientific ocean drilling, the International Ocean Discovery Program (IODP-2), U.S. staffing demonstrates some progress in broadening of participation. Importantly, the program has not tracked key metrics and dimensions of diversity, and additional work is clearly warranted in the future. Nevertheless, the program made impactful strides toward gender balance and in promoting opportunities for students and early-career researchers. By 2020, 34 percent of the science parties were women, compared with only 12 percent in the earliest days of scientific ocean drilling.[a] And since 2020, there are more U.S. women sailing on expeditions than men (Table 2.1), with one-half of the expeditions having women chief scientists. Additionally, opportunities for students and early-career scientists on expeditions have expanded during IODP-2 (Table 2.2). As shown in Figure 2.8, U.S.-operated expeditions include as many graduate students as senior researchers. Approximately one-third of all U.S. expedition science party (i.e., team) participants are graduate students, and mentoring is an integral, highly valued aspect of the program. These demographics speak to the critical importance of a U.S.-led scientific ocean drilling program in creating, training, and sustaining a diverse and inclusive oceanographic workforce.

**TABLE 2.1** Gender Demographics for U.S. Applications and Final Expedition Science Party Participation During IODP-2

| Gender | Number (%) of Applications | Number (%) of Participants (Final Science Party) |
|---|---|---|
| Male | 401 (47.7) | 165 (45.4) |
| Female | 434 (51.7) | 197 (54.3) |
| Nonbinary | 5 (0.6) | 1 (0.3) |

NOTE: IODP-2 = International Ocean Discovery Program.
SOURCE: U.S. Science Support Program, IODP.

**TABLE 2.2** U.S. IODP Science Parties by Career Level, 2015–2023

| Career Level | Number of Participants | % of Participants |
|---|---|---|
| Senior Researcher or Professor | 115 | 31.5 |
| Early- or Mid-Career | 129 | 35.3 |
| Graduate Student | 121 | 33.2 |

NOTE: IODP-2 = International Ocean Discovery Program.
SOURCE: U.S. Science Support Program, IODP.

[a] Original data from Koppers and Coggon, 2020; updated data obtained from a presentation to the committee on August 2, 2023.

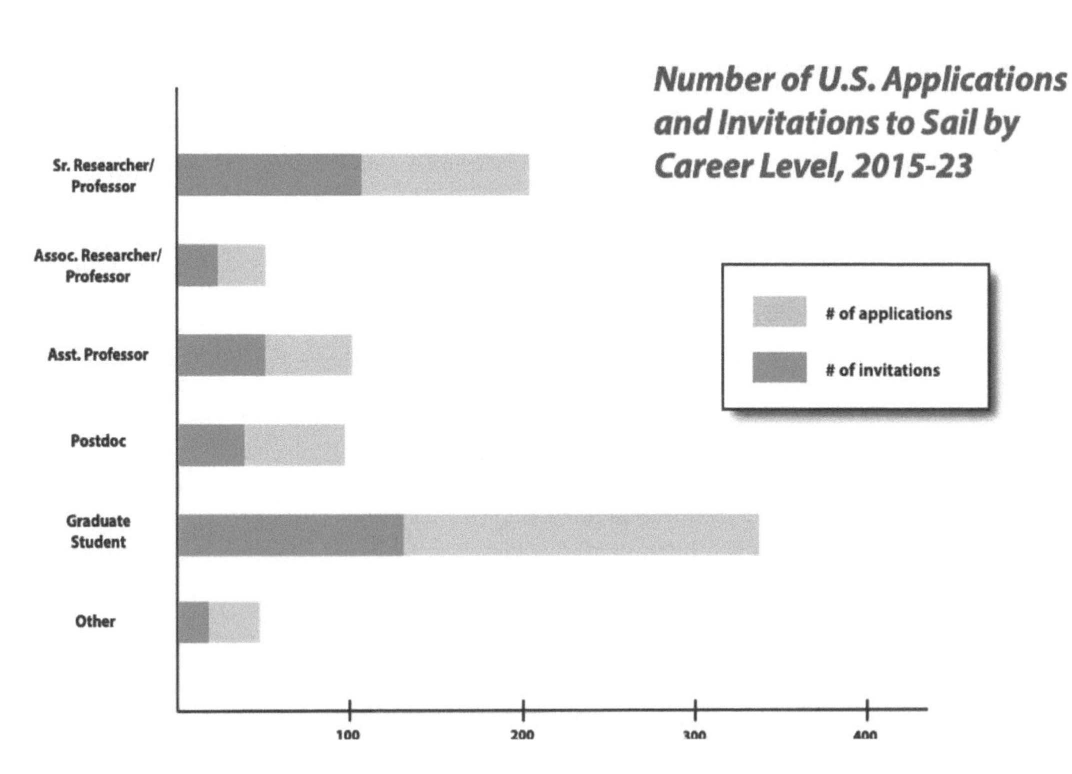


**FIGURE 2.8** Number of U.S. applications and invitations to sail on an IODP-2 expedition, listed by career level, 2015–2023. NOTE: IODP-2 = International Ocean Discovery Program.
SOURCE: U.S. Science Support Program, IODP.

## IODP-2 MANAGEMENT STRUCTURE

The current phase of the drilling program (IODP-2) depends on facilities funded by three vessel (i.e., platform) providers: NSF, the European Consortium for Ocean Research Drilling (ECORD), and Japan, with financial contributions from additional partner agencies (Australia-New Zealand Consortium, China, and India). Each platform (*JOIDES Resolution*, mission-specific platforms [MSPs], and *Chikyu*; see Table 1.1 in Chapter 1) is operated independently by the respective country or consortia, and operational decisions are guided by facility boards. The IODP-2 platforms are operated by science operators under a contract or other agreement with platform providers.[2] Science operators plan expedition logistics and manage all science and vessel operations. High-level, annual meetings serve as a venue for international discussions among funders, vessel operators, facility boards, and program member nations. Scientific activities are managed by IODP-2 program member offices.

### Overview of Expedition Science Operations

The number of expeditions per year varies among the three scientific ocean drilling platform providers and depends largely on operational costs, available funding, and time requirements for vessel mobilization and demobilization. The U.S.-operated *JOIDES Resolution,* which is dedicated entirely to scientific ocean drilling, typically operates four to five 2-month IODP expeditions per year. Consequently, the number of subseafloor holes drilled, and cores recovered, to support scientific research is much greater for the *JOIDES Resolution* than the other platforms (see Table 1.1 in Chapter 1). The drilling vessel has been the major factor in enabling U.S. leadership in the program. Although MSP expeditions provide versatile options for achieving science objectives, they do not occur as frequently as the scientific community has requested; typically, less than one expedition per year occurs (see Table 1.1 in Chapter 1). Prior to the 2023 MSP expedition (Exp 389 Hawaiian Drowned Reefs, August–October 2023), the most recent MSP expedition was in 2021. The Japan-owned and -operated *Chikyu* depends on commercial use to secure the operational funds to support scientific ocean drilling expeditions. *Chikyu* is also used for other drilling work as designated by the Japanese government and is not available to scientists during these times. The most recent IODP-2 expedition on *Chikyu* was in 2018. *Chikyu* has never conducted scientific ocean drilling outside of Japanese waters and is not expected to in the future.

Regardless of which platform is used for a scientific ocean drilling program expedition, and regardless of the specific science objectives of an expedition, there is a *generalized* workflow for preexpedition planning, the at-sea expedition, and postexpedition activities (Table 2.3). This workflow, based on decades of experience, is enabled by global science support offices that include operational, technical, curatorial, and publications staff.

Science conducted at sea depends on laboratories available on the vessel. Onboard *JOIDES Resolution* and *Chikyu,* highly instrumented shipboard laboratories exist for analysis of physical properties, downhole logging (i.e., petrophysics), micropaleontology, sediment description, petrology/structural geology, paleomagnetism, organic and inorganic geochemistry, and microbiology. Scientists conducting projects on an MSP may utilize container laboratories onboard the vessel and/or conduct only limited at-sea measurements of ephemeral properties, waiting to conduct the primary analysis of core descriptions and measurements at shore-based laboratories associated with a repository. The Bremen Core Repository (MARUM, n.d.a) and associated laboratories at the Center for Marine Environmental Sciences (i.e., MARUM)[3] at the University of Bremen in Germany meet this purpose for many MSP expeditions. In contrast, the U.S.-based Gulf Coast Repository in College Station, Texas, has limited shore-based laboratory infrastructure (e.g., XRF core scanner), as most U.S.-based laboratories are shipboard (i.e., on the *JOIDES Resolution*).[4]

[2] Through funding from NSF, the *JOIDES Resolution* Science Operator (JRSO) at Texas A&M University manages science operations of the *JOIDES Resolution.* The vessel is operated by Siem Offshore AS and is owned by Overseas Drilling Ltd. The JOIDES Resolution Facility Board provides guidance to JRSO.

[3] See https://www.marum.de/en/Infrastructure/Lab-infrastructure-at-MARUM.html.

[4] See https://iodp.tamu.edu/labs/index.html.

**TABLE 2.3** Generalized Pre- Through Postexpedition Time Frames and Activities

| Preexpedition (3–10 years) | Expedition at Sea (~2 months) | Postexpedition (1–5 years) |
|---|---|---|
| • Bottom-up proposal generation. Often via collaborative idea-generating workshops.<br>• Site surveys to characterize the subsurface geology and bathymetry.<br>• Proposal peer review via panels and committees.<br>• Vessel scheduling.<br>• Selection of expedition co-chief scientists (typically from different countries); write expectation prospectus.<br>• Application and selection of science teams. Balancing career stages, expertise, country quotas for berths.<br>• Scientists write research plans for their post-expedition research.<br>• Operations and science planning (e.g., development of contingency plans, engineering of hardware)<br>• Vessel positioning and mobilization. | • 2 months intensive collaboration and cooperation.<br>• 24/7 operations. Scientists, crew, and marine technicians all work 12-hour shifts.<br>• ~26 scientists from program member countries, if full suite of standard measurements/core descriptions are made shipboard. A smaller sea-based science team if only ephemeral measurements/ cores not opened at sea.<br>• Standard measurements of (at minimum) ephemeral properties; initial description, characterization of cores/logs.<br>• Write initial reports in near real time.<br>• Mentoring and collaborations within and across disciplines—all toward shared science objectives.<br>• Ship-to-shore public outreach.<br>• Preliminary postcruise planning with ship- and shore-based colleagues based on actual recovery. | • Curation of cores/samples/data at repository.<br>• Sampling for collaborative research addressing science objectives.<br>• Postexpedition research in laboratories around the world. Analyses are wide ranging, including geological, geochemical sedimentological, paleontological, petrographic, geophysical, paleomagnetic, microbiological, and more.<br>• Report writing/editing.<br>• Manuscript writing/editing.<br>• Collaborations, publications, presentations, and mentoring continue for many years.<br>• Additional public outreach.<br>• Collaborative grant writing and initiation of scientific ventures for future expeditions.<br>• Development of synthesis reports if part of larger regional/global initiative (e.g., Asian Monsoon). |

## HISTORY OF OCEAN DRILLING

Inspired by Project Mohole in the 1950s–1960s, scientific ocean drilling commenced operations in 1968, utilizing a dedicated drillship, the Glomar *Challenger*, as the Deep Sea Drilling Project (DSDP) (see Figure 1.1 in Chapter 1). The DSDP was a U.S.-only program during its earliest phases, with international collaborations and partnerships evolving so that it became truly global in scope—in the science conducted, the science teams involved, and the operational and management structures. With the capability to core and recover sediments and basalts at abyssal water depths (>6 km), the Glomar *Challenger* provided key data for constraining the age of the seafloor, which was primary evidence of plate tectonic theory. In addition, the recovery of overlying sediments, and subsequent analyses, yielded critical insight into Earth's history, including past changes in climate and the ocean environment over the last 150 million years. These early scientific milestones were facilitated in large part by the development of innovative drilling technologies, such as reentry cones (to replace worn core bits) and hydraulic piston coring (to recover undisturbed sequences of unconsolidated seafloor sediments that were required for constructing high-fidelity records of the past; Box 2.1). Recognizing the unique potential of the *Challenger*, in 1975, five additional nations partnered with the United States to support DSDP operations through 1983, when the *Challenger* was officially retired. By that time, the DSDP had completed 96 expeditions, having visited 624 drill sites, the findings from which revolutionized the field of Earth sciences by (1) resolving questions on the formation of Earth's continents and ocean, (2) characterizing the general composition and distribution of marine sediments globally, and (3) providing the first detailed records of ocean history (climate, circulation, chemistry, and biota) that would serve as a foundation for all future research. Textbooks in all the fundamental STEM (science, technology, engineering, and mathematics) fields had to be rewritten because of groundbreaking discoveries enabled by scientific ocean drilling.

### Ocean Drilling Program

Motivated by the outcomes of the DSDP, the international Earth sciences community embarked on a new venture, the Ocean Drilling Program (ODP) (18 partners initially), which began operations in 1985 with the drillship *JOIDES Resolution*. A larger vessel equipped with advanced dynamic positioning and coring capabilities, including active/passive heave compensation, the *JOIDES Resolution* could operate over a greater expanse of the ocean, including both shallow coastal zones and subpolar seas. Moreover, the *JOIDES Resolution* was capable of reentering previously drilled holes and installing the first subseafloor monitoring systems. Perhaps most important, with a more powerful drive system and advanced coring capabilities, core recovery rates and percentages of success in both hard and soft rock formations were significantly improved. With the insight into Earth processes gained from the DSDP, objectives of the new program were wider ranging, spanning issues from the evolution of the crust to climate to biogeochemical cycles. Although the scientific programming was initially guided by selected proposals, in order to address new initiatives and coordinate drilling expeditions, several program planning groups were created based on specific, high-priority, emerging themes (e.g., gas hydrates, extreme climates, the Arctic's role in global change, hydrogeology, seismogenic zones, deep biosphere). With guidance from the program planning groups and the *JOIDES Resolution*'s expanded coring capabilities, 110 expeditions were completed across much of the ocean through 2004, the findings of which transformed understanding of key aspects of Earth's dynamics and evolution.

### Integrated Ocean Drilling Program

In 2003, the ODP transitioned to the Integrated Ocean Drilling Program (IODP-1), which—in addition to a refurbished *JOIDES Resolution*—would begin to conduct scientific ocean drilling on other platforms. This program initially included partners Japan and ECORD, yet would eventually grow to include 17 European nations, Canada, and eventually China and South Korea as associate members. ECORD would fund and manage the MSPs, while Japan would eventually build and manage the *Chikyu*, a riser-equipped platform (see Box 2.1) with ultradeep drilling capability. The IODP-1 was a 10-year program guided by an initial science plan (IODP, 2001) that defined a set of primary objectives grouped under the broad themes of climate change, geodynamics and deep Earth cycles, and the deep biosphere. With the flexibility of multiple platforms, coring operations were extended to regions inaccessible by the *JOIDES Resolution*, such as the Arctic or western Pacific subduction zones.

### International Ocean Discovery Program

The current phase of scientific ocean drilling, the IODP-2, was launched in 2013, with the *JOIDES Resolution* as the primary platform, along with the *Chikyu* and MSPs. The program was initially supported by 21 nations, including the three major platform providers (United States, Japan, and ECORD), with additional financial contributions from five other partner agencies. The primary objectives of IODP-2 were organized into 14 "challenges" within four general themes: *climate and ocean change*, *biosphere frontiers* (i.e., subseafloor life), *Earth connections and deep processes* (i.e., ocean basin tectonics), and *Earth in motion* (i.e., geohazards) (IODP, 2011). While much of the research from this phase is still in progress, the findings of several early expeditions have already yielded significant contributions. Highlights of major accomplishments in the early years of scientific ocean drilling (DSDP and IODP-1) are described below; Chapter 3 provides an overview of accomplishments during the current phase of the program (IODP-2).

## Major Accomplishments (~1969–2013)

Since the inception of DSDP, scientific ocean drilling expeditions and research have fundamentally transformed the understanding of the planet by revealing the critical features of Earth's dynamic history, processes, and structure, including the solid Earth (i.e., upper mantle/crust), ocean, atmosphere, and ecosystems. The list of scientific ocean drilling–related achievements is extensive, represented in large part not only by the number of publications, but also, until recently, by the sustained support for drilling by the international Earth sciences community. While summarizing the full list of achievements here is impossible, notable achievements have been grouped by themes.

*Plate Tectonics and Mantle and Crustal Dynamics*

Scientific ocean drilling first tested and confirmed the theory of plate tectonics, which revolutionized geological sciences in the late 20th century. This became the basis for a new generation of models on the evolution and dynamics of Earth's crust and for understanding the origin of seismic activity and associated hazards along plate boundaries.

Notable advances resulting from scientific ocean drilling on crustal evolution and dynamics have been wide ranging. For example, the first samples of intact volcanic crust below thick layers of marine sediment were collected, revealing the complexity of crustal construction processes. Shallow sampling of large igneous provinces, which are vast outpourings of lava that have had a catastrophic influence on Earth's climate and serve as windows into deeper Earth processes (e.g., mantle convection and hot spots), took place. The scientific understanding of continental breakup, faulting, rifting, and associated magmatism was revolutionized. Additionally, the first subseafloor borehole observatory systems were installed, generating long-term data records to further understand remote environments and processes. This advancement allowed for assessments of the types of materials that are recycled by subducting plates at convergent margins. Such data illuminated fault zone behavior and related tectonic processes at active plate boundaries where Earth's largest earthquakes and tsunamis are generated. Data also revealed a subseafloor component to the ocean: large flows of fluids through virtually all parts of the seafloor, from midocean ridges to deep-sea trenches.

*Climate Evolution and Forcing*

Scientific ocean drilling extended the marine sedimentary record from the present day back to nearly 200 million years ago, allowing for reconstruction of long-term changes in global climate, including over the last 53 million years, when the planet transitioned from being hot and ice free (a greenhouse climate state) to cold and glaciated at both poles (an icehouse climate state; Figure 2.1). With advances in coring technology and strategies, much of this transition has been detailed at orbital-scale resolution (i.e., on the 10,000- to 100,000-year timescales for which changes in Earth's tilt and orbit around the sun vary significantly), which is key to assessing rates of change, as well as the presence of the climate's transient extremes. These core archives have also provided the basis for reconstructing the long-term evolution of atmospheric carbon dioxide ($CO_2$). Together with advances in numerical models, these climatic reconstructions have provided the basis for testing climate theory and establishing the sensitivity of Earth's climate (i.e., how Earth's temperature responds) to changing greenhouse gas levels, including the nature and strength of climate feedbacks. All these findings have been vital to confirming the tipping points that triggered rapid climate change in the past.

Notable contributions from scientific ocean drilling pertaining to climate evolution include both regional and global insights. For example, extensive layers of salt deposits collected deep below the seafloor verified that the Mediterranean Sea dried out repeatedly in the past, and Cretaceous black shales recovered in all ocean basins provided direct evidence of ocean-wide anoxic episodes. Collected cores have also captured the variability of key components of the global hydroclimate, including the Indian and Southeast Asian monsoons and the north–south migration of the intertropical convergence zone over the last 20,000 years and beyond. Scientific ocean drilling cores have also revealed the increased strength and permanency of El Niño–Southern Oscillation conditions during warmer global climate intervals, such as during the early Pliocene (~3–5 million years ago [Ma]).

The combined use of paleomagnetic records, radiometric dating, and the layering of marine microfossils from scientific ocean drilling expeditions from around the world were used to refine the geological timescale. Most recently, the development of astrochronology based on Earth's orbital rhythms as represented by sedimentary lithologic and geochemical cycles—documented with cores recovered by scientific ocean drilling combined with methods of absolute dating—have revolutionized paleoclimatology and paleoceanography. Moreover, these high-fidelity records were also utilized by astronomers to test and refine their models of planetary motions over tens of millions of years (e.g., Varadi et al., 2003).

Scientific ocean drilling has contributed greatly to understanding the dynamics and interconnected nature of the cryosphere, atmosphere, biosphere, and ocean in Earth's climate system. Cores from the high latitudes established the timing of the initiation of both Antarctic and Northern Hemisphere glaciations, the appearance of Arctic sea ice, and the role of greenhouse gas forcing. In particular, analyses of oxygen isotopes from microfossils in recovered cores demonstrated that the abrupt appearance of a massive ice sheet on East Antarctica approximately 34 Ma occurred after a long-term decline in atmospheric $CO_2$. Similarly, scientific ocean drilling established major changes in the character of meridional overturning circulation during the recent glacial and interglacial cycles, which were linked in part to major discharges of glacial ice. Data from ocean drilling also provided evidence of large-scale reversals of deep circulation during extreme warming events (i.e., hyperthermals) in the Eocene and evidence of a poleward shift in the westerlies during the Pliocene warm period, thus supporting forecasts for similar shifts with future global warming.

Scientific ocean drilling allowed the construction of a 100-million-year (myr) history of global sea level change, which unveiled how quickly ice sheets can melt. Data from marine sediments, in combination with geophysical models, demonstrated how sea level rise (and fall) varied from region to region. Marine sediments have also provided the basis for reconstructing the pre–ice core history[5] of atmospheric $CO_2$ extending back into the Cretaceous (>100 Ma) and testing theories on the controlling mechanisms of climate change, including feedbacks. In particular, the discovery of short-lived carbon emission events verified a theory on the role of rock weathering as a major but slow process for sequestering $CO_2$.

*Life Evolution (Marine and Continental)*

Sedimentary sample archives have been used to establish the long-term evolution and diversification of major marine zoo- and phytoplankton groups, and the influence of major climatic and environmental perturbations on marine ecosystems in extreme detail, including periods of volcanic outgassing and extreme warming, ocean acidification, and meteorite impacts.

Scientific ocean drilling has made several notable contributions on the study of life evolution. For example, work on sediment records provided the data necessary to document global patterns of plankton extinction and recovery from the Cretaceous–Paleogene impact 66 Ma, along with restoration of the marine biological pump revealed from marine sediment cores. The role of extreme warming on the rapid diversification of land mammals 56 Ma was revealed through the climate records found in cores from the deep sea. Moreover, records spanning the last 55 myr demonstrate how the efficacy of the ocean biological pump strengthened and diversity of plankton increased with long-term global cooling. In addition, scientific ocean drilling research documented the transition in the African climate relative to global cooling and the onset of Northern Hemisphere glaciation over the last 7.5 myr; the transition's influence on the evolution of mammals—including hominids—was more tightly coupled to climate change (e.g., increasing aridity) based on the study of marine sediment records.

*The Subseafloor Biosphere*

Coring of the ocean crust enabled access to some of Earth's most challenging and extreme environments, collecting data and samples of sediment, rock, fluids, and living organisms below the seafloor. A previously unknown microbial biome existing within the ocean sediments, as deep as 1.6 km below the seafloor, and within the volcanic carapace of the oceanic crust was discovered. This biome is surprisingly large and diverse, harboring new varieties of archaea, bacteria, eukarya, and viruses and varying in lithology, temperature, and redox conditions, and is likely contributing to geochemical reactions within the ocean crust, thus influencing ocean chemistry. Focus on the subseafloor, or deep biosphere, as a research theme of IODP has led to major understandings of the energetics of life, with particular emphasis on "extreme environments" (e.g., high and low temperatures, salinity

[5] Ice cores can provide detailed records of climate change ~800,000 years ago.

extremes, extremely low energy levels). As such, significant portions of the postexpedition analysis in this field were supported by the National Aeronautics and Space Administration, as the work correlates to astrobiology and the potential for non-Earth-based life.

**CONCLUSION 2.1** Research supported by ocean drilling has fundamentally transformed the understanding of the planet, making key scientific contributions to knowledge of plate tectonics, the formation and destruction of ocean crust and how these processes generate geohazards, extreme greenhouse and icehouse climates ranging across 100 myr, and the response and recovery of the biosphere to major environmental perturbations. Additionally, such research has led to the discovery of a microbial ecosystem in the environment of ocean sediments, rocks, and fluids deep below the seafloor.

# 3

# High-Priority Science Areas: Progress and Future Needs

Identifying and prioritizing critical research in ocean sciences that can be advanced only with scientific ocean drilling was an extensive process. The committee gathered community input on priorities, reviewed relevant reports that identify past priorities for scientific ocean drilling, examined progress made thus far, and identified unanswered questions that remain a priority. This chapter begins by evaluating progress made in scientific ocean drilling over the last decade with respect to priorities laid out in previous reports. It then presents the committee's framing of research priorities, and within that framing, examines progress made over the last decade specific to advancing each of the priority areas and identifies future research needs. The chapter ends by classifying scientific ocean drilling research priorities in terms of vital and urgent research and discussing how those needs fall within the national agenda.

## EVALUATING PROGRESS MADE OVER THE LAST DECADE

The committee focused on the progress made during the current funding phase of the drilling program, the International Ocean Discovery Program (IODP-2), the research of which is guided by the community-developed *2013–2023 IODP Science Plan* (IODP, 2011), organized around four science themes. From 2014 to 2023, the IODP-2 completed 57 expeditions: 46 with the *JOIDES Resolution*, 5 with the *Chikyu*, and 6 with mission-specific platforms (MSPs) (see Table 1.1 in Chapter 1). This high use of the *JOIDES Resolution,* compared with the other IODP-2 components, reflects the scientific interest and impact of the U.S.-sponsored program and the capabilities of its staff and assets.[1] IODP-2 expeditions have aimed to address each of the four science themes through their project goals (Figure 3.1), with the greatest number of expeditions addressing challenges related to *climate and ocean change*, followed by *Earth connections*. Addressing these objectives required globally ranging scientific ocean drilling capabilities, as well as specialized platforms. Drilling during IODP-2 expeditions occurred in the Atlantic, Pacific, and Indian oceans, addressing all four themes, and in the Southern Ocean, addressing the themes of *climate and ocean change* and *biosphere frontiers* (Figure 3.2).[2]

[1] While no IODP-2 expeditions were canceled due to the COVID-19 pandemic, 14 occurred after its onset, several of which were delayed and thus lost days or weeks of operations, impacting the ability to achieve all objectives.

[2] For an example of the type of reports produced by IODP expeditions, see http://publications.iodp.org/proceedings/385/385title.html. One Arctic MSP expedition was planned (Expedition 377) but had to be canceled due to Russia's attack on Ukraine.

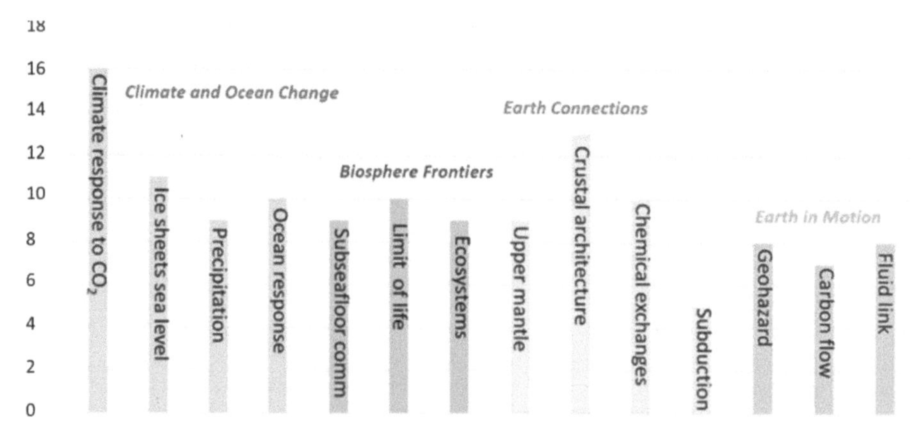


**FIGURE 3.1** Number of completed and planned expeditions during IODP-2, according to themes and challenges from the *2013–2023 IODP Science Plan*. NOTES: Comm = communities; IODP = International Ocean Discovery Program.
SOURCE: Michiko Yamamoto, IODP Science Support Office (Scripps Institution of Oceanography, University of California, San Diego).

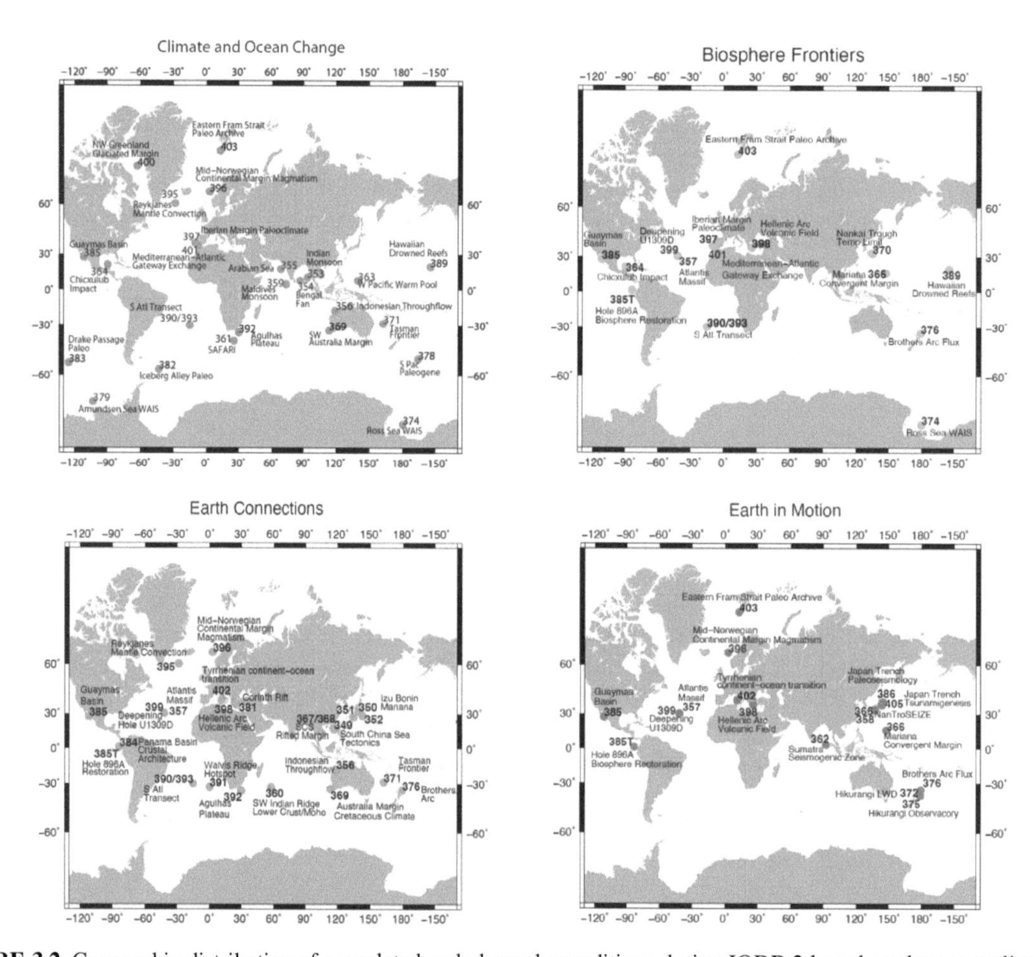


**FIGURE 3.2** Geographic distribution of completed and planned expeditions during IODP-2 based on themes outlined in the *2013–2023 IODP Science Plan*. NOTES: IODP = International Ocean Discovery Program.
SOURCE: Michiko Yamamoto, IODP Science Support Office (Scripps Institution of Oceanography, University of California San Diego).

In addition to the scientific themes and challenges posed in the *2013–2023 IODP Science Plan*, five high-priority science questions that depend on scientific ocean drilling were identified in *Sea Change: 2015-2025 Decadal Survey of Ocean Sciences* (DSOS-1) (NRC, 2015). Furthermore, the *2050 Science Framework* (Koppers and Coggon, 2020) documents the international scientific ocean drilling community's consensus on priority science areas (i.e., flagship initiatives) over the next 25 years. These flagship initiatives (see Box 2.1 in Chapter 2) are broadly similar to the *2013–2023 IODP Science Plan,* but the *2050 Science Framework* was written to address more intentionally the interconnectedness of Earth system processes in both curiosity-driven research to explore and understand and use-inspired basic research to inform and address challenges. The evolution in this framing reflects the growing interest and need in science to work more collaboratively across disciplines toward solving the complex, pressing issues facing society.

Table 3.1 provides an organizing framework to view the progress made on priorities laid out in DSOS-1 and the *2013–2023 IODP Science Plan*, aligning the DSOS-1 priorities for which scientific ocean drilling was identified as either critical or important with the *2013–2023 IODP Science Plan* challenges and themes. It lists completed IODP-2 expeditions that contributed (and continue to contribute) research outcomes addressing the prioritized science objectives and includes selected examples of key contributions toward *IODP-2 Science Plan* challenges and DSOS-1 priorities. A full review of the contributions of each expedition is not the intent of the table and is beyond the scope of this report.

Table 3.1 and this report overall are informed by expedition reports and publications of high scholarly impact. Presentations made at the DSOS-2 August 2023 meeting and the 2019 PROCEED workshop (IODP, n.d.j) hosted by the European Consortium for Ocean Research Drilling were also informative, as was *Oceanography Magazine's Special Issue on Scientific Ocean Drilling* (Kappel, 2019), which acknowledges the scientific accomplishments and evolution of the drilling program over its 50-year history. However, it is also worth noting what is not informing this synthesis: to the knowledge of the committee, **the scientific ocean drilling program has not conducted a formal evaluation of progress made toward the identified *Science Plan* challenges during the current funding phase**. The lack of documented assessment—by the IODP Forum, IODP science facility boards or committees, and/or IODP science operators—on how the program is progressing toward addressing its specific priorities is a weakness in the program and contrasts with its pattern of regular operational assessments.

> **CONCLUSION 3.1** The scientific ocean drilling program would benefit from developing and executing a formal evaluation for assessing progress made toward achieving scientific priorities and for communicating and sharing the program's achievements and value.

**While much of the research from the current phase of the program is still in progress, the findings from several expeditions have already yielded significant contributions**. Highlights of some of the scientific accomplishments identified in Table 3.1, as well as other key accomplishments, are described further in the sections that follow, organized within five high-priority research areas identified by the committee. This is not an all-encompassing review of the full scope of contributions, nor does it include unexpected discoveries that occurred outside of the prescribed boundaries of the *Science Plan* challenges or DSOS-1 priorities. Notably, accomplishments in science do not occur in isolation. They are often enhanced by connections to other fields of study (see Box 1.1 in Chapter 1), advances in tools and technologies (Box 3.1), development of new analytical methods and proxies (Box 3.2), and support provided by a diverse workforce (see Box 2.2 in Chapter 2).

**TABLE 3.1** Progress Made Toward Past Research Priorities

| | **Completed IODP-2 Expeditions That Contributed or Are Contributing to Priority Areas** | **Selected Examples of Key Contributions** |
|---|---|---|
| **IODP-2 Theme: Climate and Ocean Change: Reading the Past, Informing the Future** | | |
| **DSOS-1 What are the rates, mechanisms, impacts, and geographic variability of sea level change?**<br>**IODP-2: How do ice sheets and sea level respond to a warming climate?** | Exp 359 Maldives Monsoon and Sea Level<br>Exp 369 Australia Cretaceous Climate and Tectonics<br>Exp 374 Ross Sea West Antarctic Ice Sheet<br>Exp 379 Amundsen Sea West Antarctic Ice Sheet<br>Exp 382 Iceberg Alley and Subantarctic Ice and Ocean Dynamics<br>Exp 383 Dynamics of Pacific Antarctic Circumpolar Current<br>Exp 389* Hawaiian Drowned Reefs<br>Exp 390/393 South Atlantic Transect<br>Exp 400 NW Greenland Glaciated Margin | **Documented influence of ice sheet dynamics on the magnitude of sea level change.**<br>• Determined a larger-than-present West Antarctic Ice Sheet in the early middle Miocene that explains very large global sea level amplitudes previously documented for that period.<br>• Documented evidence in the Amundsen and Ross seas of a highly unstable West Antarctic Ice Sheet coinciding with the early Pliocene warm period between 4.2 and 3.2 Ma.<br>• Demonstrated over the last 1 myr, Northern Hemisphere ice sheet expansion occurred during times of declining obliquity, whereas times of contraction were tied to minima in precession, in contrast to pre 1 Ma when obliquity was the dominant control on ice volume. |
| **DSOS-1: How have ocean biogeochemical and physical processes contributed to today's climate and its variability, and how will this system change over the next century?**<br>**IODP-2: How does Earth's climate system respond to elevated levels of atmospheric $CO_2$?** | Exp 361 South African Climates<br>Exp 363 Western Pacific Warm Pool<br>Exp 369 Australia Cretaceous Climate and Tectonics<br>Exp 371 Tasman Frontier Subduction Initiation & Paleogene Climate<br>Exp 378 South Pacific Paleogene Climate<br>Exp 382 Iceberg Alley and Subantarctic Ice and Ocean Dynamics<br>Exp 383 Dynamics of Pacific Antarctic Circumpolar Current<br>Exp 392 Agulhas Plateau Cretaceous Climate<br>Exp 390/393 South Atlantic Transect<br>Exp 395 Reykjanes Mantle Convection<br>Exp 396 Mid-Norwegian Continental Margin Magmatism<br>Exp 397 Iberian Margin Paleoclimate | **Gathered data on ocean circulation and climate sensitivity to changing greenhouse gas levels.**<br>• Determined that Indian Ocean salinity buildup during glacials impacted deglacial circulation recovery via the Agulhas Leakage.<br>• Documented a tropical sea surface warming trend over the last 12 kyr consistent with climate models.<br>• Documented the northward shift in Antarctic icebergs and sea ice melt, key to Atlantic meridional overturning circulation reorganization during glacials.<br>• Provided first evidence of middle Eocene climate optimum global warming at abyssal depths; shows that acidification affected the entire oceanic water column during this event. |

*continued*

**TABLE 3.1** Continued

| | Completed IODP-2 Expeditions That Contributed or Are Contributing to Priority Areas | Selected Examples of Key Contributions |
|---|---|---|
| **IODP-2: What controls regional patterns of precipitation, such as those associated with monsoons or El Niño?** | Exp 353 Indian Monsoon Rainfall<br>Exp 354 Bengal Fan<br>Exp 355 Arabian Sea Monsoon<br>Exp 356 Indonesian Throughflow<br>Exp 359 Maldives Monsoon and Sea Level<br>Exp 361 South African Climates<br>Exp 363 Western Pacific Warm Pool<br>Exp 389* Hawaiian Drowned Reefs | **Made major progress in understanding regional monsoon precipitation.**<br>• Validated model-predicted increased monsoon precipitation and extreme variability due to greenhouse gas forcing with reconstruction of Pleistocene summer monsoon rainfall record.<br>• Constrained the past evolution (initiation and strengthening) of the South Asian monsoon with cores from the Arabian Sea.<br>• Documented weakening of the Indonesian throughflow (ITF) at 1.55 and 0.65 Ma, coinciding with ice sheet expansion, sea level change, and drying of western Australia, suggesting that restrictions of the ITF influenced both the evolution of global ocean circulation and the development of the modern hydrological cycle.<br>• Determined that high-latitude cooling around Antarctica in the Miocene drove changes in precipitation patterns in Australia and Southeast Asia from 12–8 Ma. |
| **IODP-2: How resilient is the ocean to chemical perturbations?** | Exp 364* Chicxulub K-T Impact Crater<br>Exp 369 Australia Cretaceous Climate and Tectonics<br>Exp 378 South Pacific Paleogene Climate<br>Exp 392 Agulhas Plateau Cretaceous Climate<br>Exp 390/393 South Atlantic Transect<br>Exp 396 Mid-Norwegian Continental Margin Magmatism | **Identified physical and biogeochemical changes that affect ecosystems and climate.**<br>• Documented massive volcanic eruptions that triggered widespread ocean acidification and ecological stress in the middle Cretaceous.<br>• Determined that microbial blooms triggered rapid precipitation of calcite in the vicinity of crater site immediately following the impact.<br>• Agulhas Plateau drilling retrieved a sedimentary record of enhanced basalt weathering that provides a natural laboratory for investigating the impacts of proposed climate mitigation techniques. |
| **IODP-2 Theme: Earth Connections: Deep Processes and Their Impact on Earth's Surface Environment** | | |
| **DSOS-1: What are the processes that control the formation and evolution of ocean basins?**<br><br>**IODP-2: What are the composition, structure, and dynamics of Earth's upper mantle?** | Exp 356 Indonesian Throughflow<br>Exp 357* Atlantis Massif Seafloor Processes: Serpentinization and Life<br>Exp 360 SW Indian Ridge Lower Crust and Moho<br>Exp 384 *JOIDES Resolution* Engineering Testing<br>Exp 391 Walvis Ridge Hotspot<br>Exp 392 Agulhas Plateau Cretaceous Climate<br>Exp 395/395C Reykjanes Mantle Convection and Climate<br>Exp 396 Mid-Norwegian Margin Magmatism<br>Exp 398 Hellenic Arc Volcanic Field<br>Exp 399 Building Blocks of Life, Atlantis Massif | **Fulfilled a 60-year goal of scientific ocean drilling by drilling into upper mantle rock.**<br>• Conducted engineering tests with the goal of improving the chances of success in deep (>1 km) drilling and coring in igneous ocean crust.<br>• Drilled 1.5 km into Earth's upper mantle at a "tectonic window" of the Atlantis Massif, along the midocean ridge. |

*continued*

**TABLE 3.1** Continued

| | Completed IODP-2 Expeditions That Contributed or Are Contributing to Priority Areas | Selected Examples of Key Contributions |
|---|---|---|
| **IODP-2 Theme: Earth Connections: Deep Processes and Their Impact on Earth's Surface Environment** | | |
| **IODP-2: How are seafloor spreading and mantle melting linked to ocean crustal architecture?** | Exp 349 South China Sea Tectonics<br>Exp 360 SW Indian Ridge Lower Crust and Moho<br>Exp 366 Mariana Convergent Margin<br>Exp 367/368 South China Sea Rifted Margin<br>Exp 381* Corinth Active Rift Development<br>Exp 384 *JOIDES Resolution* Engineering Testing<br>Exp 385 Guaymas Basin Tectonics and Biosphere<br>Exp 385T Panama Basin Crustal Architecture and Deep Biosphere<br>Exp 390/393 South Atlantic Transect<br>Exp 391 Walvis Ridge Hotspot<br>Exp 392 Agulhas Plateau Cretaceous Climate<br>Exp 395/395C Reykjanes Mantle Convection and Climate<br>Exp 396 Mid-Norwegian Margin Magmatism | **Elucidated the processes by which ocean crustal architecture is created and modified, from rifting to seafloor spreading.**<br>• Constrained the initiation of seafloor spreading in the South China Sea to 33 Ma, with a rapid (<10 myr) transition between continental breakup and igneous seafloor spreading.<br>• Identified carbonated silicate melts, previously only predicted by experimental studies, in subseafloor of South China Sea.<br>• Quantified the tectono–magmatic interactions that form and modify the lower oceanic crust at Atlantis Bank, on the ultraslow-spreading Southwest Indian Ridge.<br>• Demonstrated the processes by which gabbros at Atlantis Bank crystallized in the lower crust and were later modified by crystal–plastic deformation and faulting. |
| **IODP-2: How do subduction zones initiate, cycle volatiles, and generate continental crust?** | Exp 350 Izu-Bonin-Mariana Rear Arc<br>Exp 351 Izu-Bonin-Mariana Arc Origins<br>Exp 352 Izu-Bonin-Mariana Forearc<br>Exp 358** NanTroSEIZE: Plate Boundary Deep Riser<br>Exp 365** NanTroSEIZE: Shallow Megasplay Long-Term Borehole<br>Exp 366 Mariana Convergent Margin<br>Exp 371 Tasman Frontier Subduction Initiation and Paleogene Climate<br>Exp 375 Hikurangi Subduction Margin<br>Exp 380** NanTroSEIZE: Frontal Thrust Borehole Monitoring System<br>Exp 398 Hellenic Arc Volcanic Field | **Used drilling results to understand how mantle melting processes evolve during and after subduction initiation.**<br>• Recovered a submarine sedimentary record of magmatic arc history from birth to demise.<br>• Documented evidence of spontaneous subduction initiation based on basement rocks formed during inception of the Izu–Bonin–Mariana subduction system.<br>• Identified complex, far-field uplift and depression accompanying the inception of the Tonga–Kermadec subduction system, which may have involved both spontaneous and induced elements.<br>• Cores from Hikurangi demonstrated that slow slip events and associated slow earthquake phenomena are promoted by lithological, mechanical, and frictional heterogeneity within the fault zone, enhanced by geometric complexity associated with subduction of rough crust. |

*continued*

**TABLE 3.1** Continued

| | **Completed IODP-2 Expeditions That Contributed or Are Contributing to Priority Areas** | **Selected Examples of Key Contributions** |
|---|---|---|
| **DSOS-1: What is the geophysical, chemical, and biological character of the subseafloor environment**<br>**IODP-2: What are the mechanisms, magnitude, and history of chemical exchanges between the oceanic crust and seawater?** | Exp 357* Atlantis Massif Seafloor Processes: Serpentinization and Life<br>Exp 366 Mariana Convergent Margin<br>Exp 376 Brothers Arc Flux<br>Exp 385 Guaymas Basin Tectonics and Biosphere<br>Exp 385T Panama Basin Crustal Architecture and Deep Biosphere<br>Exp 390/393 South Atlantic Transect<br>Exp 392 Agulhas Plateau Cretaceous Climate<br>Exp 395/395C Reykjanes Mantle Convection and Climate | **Developed new insights and models for chemical and fluid exchanges between ocean crust and seawater.**<br>• Determined anomalies in the marine silica budget that may be explained by low-temperature serpentine alteration by seawater.<br>• Developed a new model for slab dehydration and melting beneath the Mariana arc that provides fluids/melt, triggering volcanic eruptions.<br>• Provided new insights into the hydrothermal mobility and chemical exchanges of rhenium and osmium isotopes, which are powerful tools for geochronology and tracing geochemical processes. |
| **IODP-2 Theme: Biosphere Frontiers: Deep Life and Environmental Forcing of Evolution** | | |
| **DSOS-1: What is the geophysical, chemical, and biological character of the subseafloor environment and how does it affect global elemental cycles and understanding of the origin and evolution of life?**<br>**IODP-2: What are the origin, composition, and global significance of deep subseafloor communities?** | Exp 357* Atlantis Massif Seafloor Processes: Serpentinization and Life<br>Exp 366 Mariana Convergent Margin<br>Exp 376 Brothers Arc Flux<br>Exp 385 Guaymas Basin Tectonics and Biosphere<br>Exp 385T Panama Basin Crustal Architecture and Deep Biosphere<br>Exp 390/393 South Atlantic Transect<br>Exp 398 Hellenic Arc Volcanic Field | **Revealed global diversity of microbial communities in subseafloor environments.**<br>• Extended understanding of abiotic organic synthesis and diversification in hydrothermal environments, which involve magmatic degassing and water-consuming mineral reactions.<br>• Identified complex subsurface hydrothermal fluid mixing at a submarine arc volcano that supports distinct and highly diverse microbial communities. |
| **IODP-2: What are the limits of life in the subseafloor realm?** | Exp 360 Indian Ridge Lower Crust and Moho<br>Exp 370** Temperature Limit of the Deep Biosphere off Muroto<br>Exp 375 Hikurangi Subduction Margin Observatory<br>Exp 376 Brothers Arc Flux<br>Exp 385 Guaymas Basin Tectonics and Biosphere<br>Exp 385T Panama Basin Crustal Architecture and Deep Biosphere<br>Exp 390/393 South Atlantic Transect<br>Exp 398 Hellenic Arc Volcanic Field | **Made pioneering observations about microbial life in extreme environments.**<br>• Proposed that methanogenesis associated with serpentinization could support a whole new planetary biosphere deep in the oceanic basement.<br>• Documented low-biomass, diverse microbial population survival strategies in lower-crustal rocks.<br>• Documented an active methanogenic and sulfate-reducing population in deeply buried sediments (1,200 m) at temperatures up to ~120°C.<br>• Identified variation and diversity of community composition and function in a complex, hydrothermally active submarine volcano. |

*continued*

**TABLE 3.1** Continued

| | **Completed IODP-2 Expeditions That Contributed or Are Contributing to Priority Areas** | **Selected Examples of Key Contributions** |
|---|---|---|
| **IODP-2: How sensitive are ecosystems and biodiversity to environmental change?** | Exp 363 Western Pacific Warm Pool<br>Exp 364* Chicxulub K-T Impact Crater<br>Exp 382 Iceberg Alley and Subantarctic Ice and Ocean Dynamics<br>Exp 383 Dynamics of Pacific Antarctic Circumpolar Current<br>Exp 385 Guaymas Basin Tectonics and Biosphere<br>Exp 389* Hawaiian Drowned Reefs<br>Exp 390/393 South Atlantic Transect<br>Exp 397 Iberian Margin Paleoclimate<br>Exp 398 Hellenic Arc Volcanic Field | **Documented environmental changes and their ecosystem responses on a range of timescales and oceanic settings.**<br>• Demonstrated a strong temperature control on the efficacy of the biological pump and carbon cycling in the upper ocean.<br>• Determined that plankton evolution and diversity have been paced by orbitally forced changes in climate and the carbon cycle over the last several million years.<br>• Identified antiphased (i.e., alternating layers) dust deposition and biological productivity in the Antarctic Zone over 1.5 myr.<br>• Documented rapid recovery of marine benthic and planktic life at the Chicxulub impact crater. |
| **IODP-2 Theme: Earth in Motion: Processes and Hazards on Human Timescales** | | |
| **DSOS-1: What is the geophysical, chemical, and biological character of the subseafloor environment and how does it affect global elemental cycles?**<br>**IODP-2: What properties and processes govern the flow and storage of carbon in the subseafloor?** | Exp 357* Atlantis Massif Seafloor Processes: Serpentinization and Life<br>Exp 372 Creeping Gas Hydrate Slides and Hikurangi LWD (logging-while-drilling)<br>Exp 375 Hikurangi Subduction Margin Observatory<br>Exp 381* Corinth Active Rift Development<br>Exp 385 Guaymas Basin Tectonics and Biosphere<br>Exp 386* Japan Trench Paleoseismology | **Documented new carbon-cycling links between the Earth's surface and its deeper interior along plate boundaries.**<br>• Demonstrated that carbon cycling is enhanced by earthquakes and microbial mediation in hadal trench environments.<br>• Illustrated that greater carbon burial in a young tectonic rift (continental margin) setting during interglacials can be seen in marine locations rather than during glacials, when the setting was closed off from ocean. |
| **IODP-2: How do fluids link subseafloor tectonic, thermal, and biogeochemical processes?** | Exp 357* Atlantis Massif Seafloor Processes: Serpentinization and Life<br>Exp 365 **NanTroSEIZE: Shallow Megasplay Long-Term Borehole<br>Exp 366 Mariana Convergent Margin<br>Exp 370** Temperature Limit of the Deep Biosphere off Muroto<br>Exp 375 Hikurangi Subduction Margin Observatory<br>Exp 376 Brothers Arc Flux<br>Exp 380** NanTroSEIZE: Frontal Thrust Borehole Monitoring System<br>Exp 385 Guaymas Basin Tectonics and Biosphere<br>Exp 385T Panama Basin Crustal Architecture and Deep Biosphere<br>Exp 396 Mid-Norwegian Margin Magmatism | **Core records and borehole instruments advanced characterization of fluid flow in a range of environmental settings and made connections to climate change.**<br>• Allowed sampling of fluids for geochemistry and microbiology over multiple years using eight new or refurbished boreholes instrumented with observatories in environments from midocean ridges to subduction zones. [Total instrumented boreholes now ~50.]<br>• Identified a new type of intermediate-stage hydrothermal system in the Gulf of California, which serves as a critical missing link to understanding the complex tectonic, thermal, and biochemical evolution of hydrothermal systems.<br>• Linked shallow-water hydrothermal venting to an extreme global warming event, the Paleocene–Eocene thermal maximum. |

*continued*

**TABLE 3.1** Continued

| | **Completed IODP-2 Expeditions That Contributed or Are Contributing to Priority Areas** | **Selected Examples of Key Contributions** |
|---|---|---|
| **DSOS-1: How can risk be better characterized and the ability to forecast geohazards such as mega-earthquakes, tsunamis, undersea landslides, and volcanic eruptions be improved?**<br><br>**IODP-2: What mechanisms control the occurrence of destructive earthquakes, landslides, and tsunami?** | Exp 358** NanTroSEIZE: Plate Boundary Deep Riser 4<br>Exp 362 Sumatra Seismogenic Zone<br>Exp 365** NanTroSEIZE: Shallow Megasplay Long-Term Borehole<br>Exp 372 Creeping Gas Hydrate Slides and Hikurangi LWD<br>Exp 374 Ross Sea West Antarctic Ice Sheet<br>Exp 375 Hikurangi Subduction Margin Observatory<br>Exp 380** NanTroSEIZE: Frontal Thrust Borehole Monitoring System<br>Exp 381* Corinth Active Rift Development<br>Exp 386* Japan Trench Paleoseismology<br>Exp 398 Hellenic Arc Volcanic Field | **Made major progress in deep drilling of plate boundaries and understanding a range of fault types, geologic properties, and motions leading to earthquakes; and new recognition of climatically linked submarine landslides.**<br>• Borehole observatories demonstrated the potential to capture the spectrum of fault locking and strain release on short timescales.<br>• Subduction-zone borehole observatories, actively monitoring strain accumulation and release far offshore, have detected slip events and locking behavior—including triggered and spontaneous events—in the shallowest tsunamigenic reaches of the megathrust.<br>• Fault sampling demonstrated that some shallow faults preserve signals of rapid, very local heating, which can be explained only by fast slip.<br>• Determined that freshwater release from the dehydration mineral-bound water during sediment burial in thickly sedimented subduction zones plays a role in triggering strong earthquakes and tsunamis.<br>• Recognized that melting glaciers can trigger submarine mass failure (landslides), which could trigger tsunamis. |

*Conducted on a mission-specific platform.
**Conducted aboard the *Chikyu*.
NOTES: Examples of progress made during the International Ocean Discovery Program (IODP-2) are mapped against research priorities included in *Sea Change: 2015–2015 Decadal Survey of Ocean Sciences* (DSOS-1) (NRC, 2015) that require or include ocean drilling and challenges, and themes from the *2013–2023 IODP Science Plan* (IODP, 2011). Expeditions with no asterisks were conducted aboard the *JOIDES Resolution*. Scheduled_expeditions for the remainder of the current phase of the IODP program: Exp 401 Mediterranean-Atlantic Gateway Exchange; Exp 402 Tyrrhenian Continent-Ocean Transition; Exp 403 Eastern Fram Strait Paleo-archive; and Exp 405** Japan Trench_Tsunamigenesis. Exp = expedition; kyr = thousand years; Ma = million years ago; myr = million years.
SOURCE: List of categorized expeditions informed by Brinkhuis, 2023.

## BOX 3.1
## Advances in Tools and Technologies

Scientific progress can be advanced by developing and diversifying tools and technologies, as exemplified by scientific ocean drilling. Operational enhancements in response to requests by the scientific community during the International Ocean Discovery Program (IODP-2) include:

- **New options for core recovery in special settings.** Giant piston coring (see Box 2.1 in Chapter 2) is a suitable coring approach for achieving scientific objectives that require high-resolution records of the very recent past (late Pleistocene to Holocene). Giant piston coring was used for the first time on mission-specific platform (MSP) Expedition 386 (Strasser et al., 2019), which aimed to recover a continuous record of prehistoric (preinstrumental) earthquake events in the Japan Trench at over 8,000-m water depth.
- **Use of seabed drilling systems.** Seabed drilling systems (Box 2.1) were used for the first time for the microbiology- and tectonics-focused MSP Expedition 357 (Früh-Green, 2015), in order to recover a complex, shallow mantle sequence on the flank of the Mid-Atlantic Ridge. The expedition successfully utilized other new technologies, including an in situ sensor package and water-sampling system placed on the seabed drills to evaluate physical and chemical properties (e.g., dissolved oxygen, methane, pH, temperature, and conductivity) during drilling. Additionally, a borehole plug system was installed at the drill site, allowing reaccess for future sampling, which then demonstrated that contamination tracers can be delivered into drilling fluids when using seabed drills.
- **Piston coring tools to improve core recovery in challenging lithologies.** The advanced piston corer (APC) (Box 2.1) used on the *JOIDES Resolution* is the primary coring tool used to obtain the highest-quality cores for high-resolution climate and paleoceanographic studies. However, it does not work well when the subseafloor layers are too firm or when hard and soft layers alternate. To address these operational challenges, a new, shorter (4.7-m) version of the APC, called the half-length APC (HLAPC) (IODP, n.d.k) was developed. Since 2013, the HLAPC as been useful in extending the depth (i.e., age) range for recovering undisturbed sediment suitable for high-resolution research (IODP, n.d.l). The HLAPC has also been useful in recovering difficult-to-core lithologies. For example, it recovered critical intervals of sands in the Bengal and Nicobar fans (Expeditions 354 and 362), at depths up to 800 m below the seafloor, and in the Mariana serpentinite mud volcanoes (Expedition 366). The HLAPC has been used extensively during IODP-2, accounting for about 21 percent of all piston coring.
- **New drill-in-casing system and hydrologic release tool to save operational time and cost.** Deep sediment holes, including those that penetrate basement rock below sediments, traditionally require a deep hole to be predrilled and double- and triple-casing walls (referred to as casing strings) to be installed to stabilize the upper hole. These are time-consuming efforts, often requiring 7–10 days of ship time. A drill-in-casing system was developed for the *JOIDES Resolution* to save time and hardware costs when scientific objectives require deep sediment penetration or when starting holes in bare rock. The concept was demonstrated in 2014 as a more time-efficient approach to drilling in a single casing string with a special reentry system and without predrilling a hole, allowing a greater number of deep-penetration holes to be attempted and at a lower cost. In addition, a hydraulic release tool (HRT) (IODP, n.d.g) was adapted to drill in a reentry system with a short casing string to start a hole in bare-rock seafloor at Southwest Indian Ridge (Expedition 360). The HRT reentry system has continued to be simplified and is now being used as the standard drill-in-casing system to establish a single-casing string for deep sediment penetration. As of September 2023, 23 holes have been cased using this approach, collectively saving ~90–140 days of operational time and leaving time to achieve other science objectives.

**BOX 3.2**
**Analytical Chemistry and Proxies of Past Ocean Conditions**

Understanding Earth's evolution, from the genesis of ocean crust to changes in climate to the diversity of the subseafloor microbial communities, is based largely on biogeochemical evidence extracted from rocks and fossils recovered by scientific ocean drilling. The process of extracting this information, however, is challenging, hinging on the ability to analyze a wide range of materials, often on the micro scale, with high precision. As such, technological advances in analytical chemistry, from sample extraction and processing to instrumentation, have played a critical role in addressing a range of scientific questions. This critical role is particularly evident in efforts to reconstruct changes in ocean temperatures and chemistry by proxy (i.e., based on the chemical/isotopic composition of planktonic and benthic microfossils). Much of the pioneering work on reconstructing variations in ocean temperature, based on the oxygen isotope ratios ($^{18}O/^{16}O$) of calcite shells of plankton, was facilitated by the development of mass spectrometers. Further advances in mass spectrometry then allowed for the separation and analysis of carbon isotopic ratios ($^{13}C/^{12}C$) of algal organic compounds, a proxy for seawater carbon dioxide ($CO_2$) concentrations. In combination, these advances enabled the first assessment of past climate sensitivity to greenhouse gas forcing, albeit with large uncertainties. The mass spectrometer, related technologies, and analytical techniques, have continued to advance further (Figure 3.3), allowing for reduced uncertainties and the continued development of new proxies of a wide range of seawater parameters, such as temperature, salinity, dissolved $O_2$, and nutrient concentrations and pH (e.g., boron isotopes), leading to the reconciliation of past changes in climate, ocean dynamics, and biogeochemical cycling. While the new proxies are being applied to legacy materials, most critical intervals have been depleted to the point where additional cores are required to take advantage of these recent technological developments.

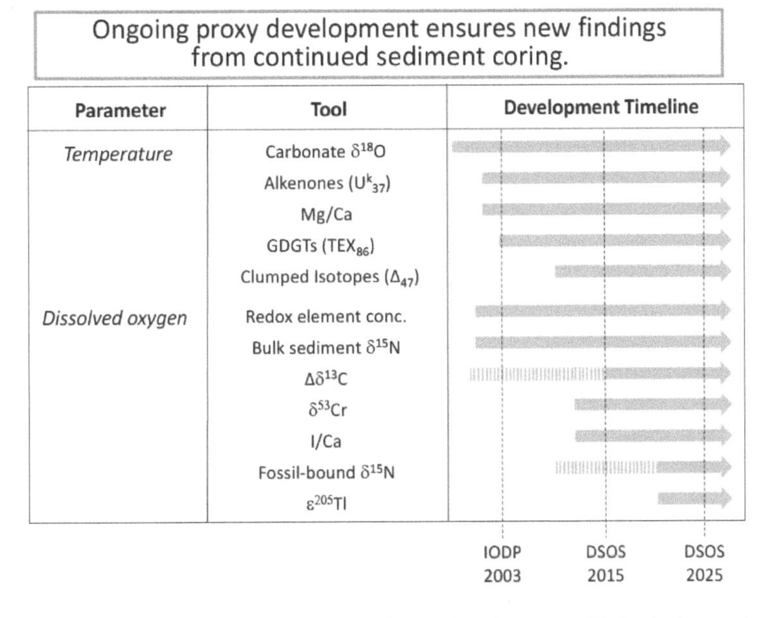


**FIGURE 3.3** Timelines of proxies of temperature and dissolved oxygen ($O_2$) relative to the phases of scientific ocean drilling. NOTES: The development and testing of several new proxies are underway, providing more tools for reconstructing these past climate and ocean conditions. DSOS = Decadal Survey of Ocean Sciences; GDGT = glycerol diakyl glycerol tetraether; IODP = International Ocean Discovery Program. SOURCE: From Jesse R. Farmer, University of Massachusetts Boston, and Daniel M. Sigman, Princeton University.

## SCIENTIFIC OCEAN DRILLING RESEARCH PRIORITIES

The remainder of this chapter is organized under the five research areas (Figure 3.4) that the committee identified as high priority and that continue to require scientific ocean drilling to be understood:

- ground truthing climate change
- evaluating marine ecosystem responses to climate and ocean change
- monitoring and assessing geohazards
- exploring the subseafloor biosphere
- characterizing the tectonic evolution of the ocean basins

The committee's five high-priority areas are informed by, but independent of, previous scientific ocean drilling planning efforts. Although details and nuances vary, these priorities are consistent with the *2050 Science Framework* flagship initiatives and priorities laid out in preceding science plans. The five high-priority areas have broad topical relationships to the scientific questions that emerged during the first DSOS-1 review and to the current IODP-2 *Science Plan* (Table 3.1), and they incorporate aspects of the *2050 Science Framework*'s strategic objectives, which highlight the research needed to understand the interconnected processes in the Earth system (Figure 1.4).

**CONCLUSION 3.2** The committee identified five (unranked) high-priority research areas that require future scientific ocean drilling: (a) ground truthing climate change, (b) evaluating past marine ecosystem responses to climate and ocean change, (c) monitoring and assessing geohazards, (d) exploring the subseafloor biosphere, and (e) characterizing the tectonic evolution of ocean basins. Though differing in detail and nuance, the priority areas align with the initiatives identified by the scientific ocean drilling community.

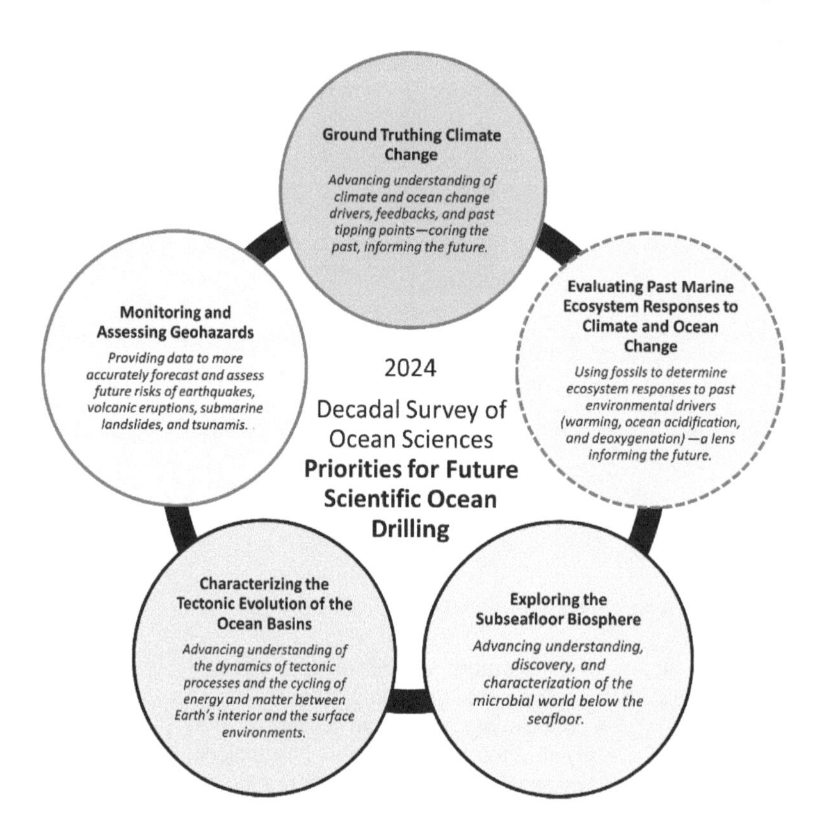


**FIGURE 3.4** Priorities for future scientific ocean drilling. NOTE: All five priority areas are considered vital; those outlined in red are also deemed urgent.

## Ground Truthing Climate Change

*Advancing understanding of climate and ocean change drivers, feedbacks, and past tipping points—coring the past, informing the future.*

Earth's climate is currently in a transient (nonequilibrium) state because of the unprecedented rate of greenhouse gas emissions in the past 100+ years. Earth's climate system has not fully responded to the dramatic increase in greenhouse gases; changes are still occurring in response to the forcing (i.e., drivers). Additionally, different parts of the climate system (e.g., ice sheets vs. sea ice, surface ocean vs. deep ocean, polar regions vs. temperate regions) are responding at different rates. As such, direct observations of the global climate from less than a century ago provide too little data to adequately assess the ability of advanced models to accurately simulate Earth's climate at greenhouse gas levels significantly higher (or lower) than present (Figure 3.5).

Primary sources of model uncertainty include feedbacks, both physical (ocean circulation, heat storage and transport, clouds) and biogeochemical (carbon cycle), that can potentially amplify (or dampen) the response to forcing. That the response of both physical and biogeochemical feedbacks is nonlinear poses a significant challenge for modeling. As such, **testing the skill of models, reducing uncertainties, and learning more about the Earth system's response to changes in forcing (i.e., greenhouse gases), requires an examination of past changes in climate as case studies.**

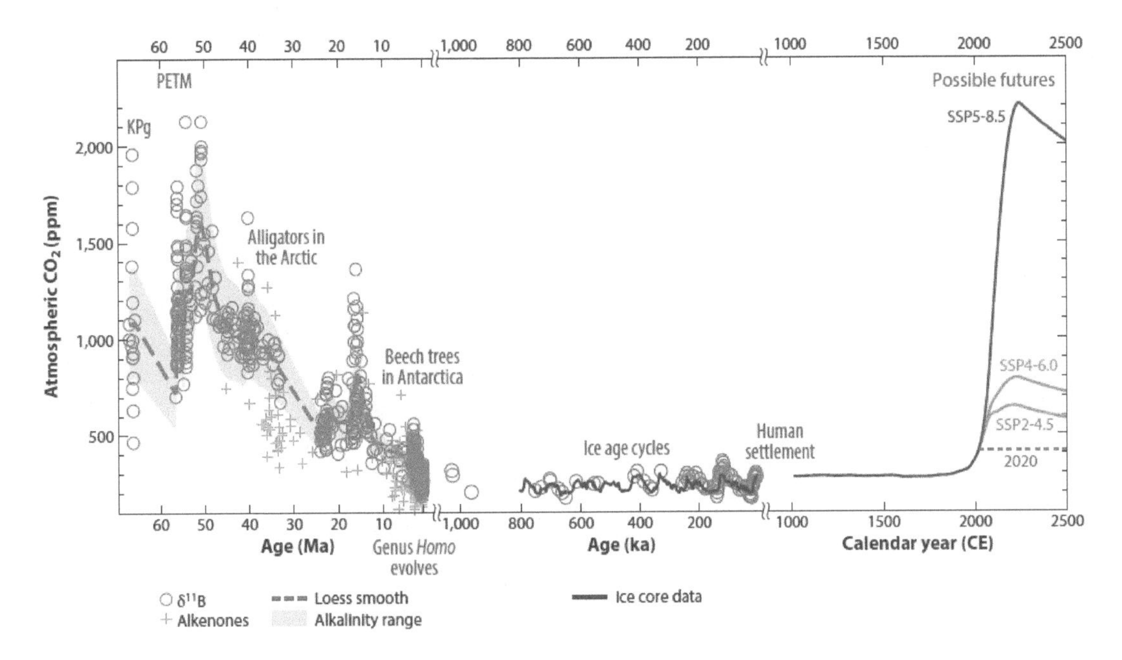


**FIGURE 3.5** Reconstruction of atmospheric carbon dioxide ($CO_2$) over the Cenozoic (0–66 million years ago [Ma]) compared with $CO_2$ scenarios associated with shared socioeconomic pathways (SSPs) SSP2-4.5, SSP4-6.0, and SSP5-8.5 (Meinshausen et al., 2020; Rae et al., 2021). The paleo $CO_2$ estimates are derived primarily from alkenones, a class of organic compounds, and the boron isotope composition of fossil plankton preserved in sediment cores recovered by the Ocean Drilling Program and Integrated Ocean Drilling Program. The chemical structure of alkenones is regulated by water temperature, which is sensitive to atmospheric $CO_2$. The boron isotope composition of fossil plankton can be used to estimate the level of ocean acidification due to $CO_2$. Note that humans (*Homo sapiens*) evolved ~200,000 years ago, when atmospheric $CO_2$ was oscillating between 180 and 280 parts per million (ppm). As such, this is the first time humans have lived under such elevated $CO_2$ conditions. PETM = Paleocene–Eocene thermal maximum; ka = thousand years ago; K–Pg = Cretaceous–Paleogene.
SOURCE: Used with permission of Rae et al., 2021.

Determining how much the global average temperature is expected to change in response to a given change in the amount of atmospheric greenhouse gases is challenging but essential to refining model forecasts of future climate scenarios. Scientific ocean drilling plays an important role in achieving this objective. Earth's equilibrium climate sensitivity to greenhouse gas (e.g., carbon dioxide [$CO_2$]) forcing has long been unresolved in climate models, exhibiting a wide range of sensitivities, from ~2.0 to 5.0°C per doubling of $CO_2$ (see Sherwood et al., 2020). Observations of past climates, particularly over long periods of time with extremes in $CO_2$ (e.g., early Eocene climatic optimum, 53 Ma), can provide insight into equilibrium climate states under a wide range of atmospheric $CO_2$ concentrations (~180–2,000 ppm) (Rohling et al., 2018) and into transient climate states when the rate of rise in greenhouse gas was on the scale of modern rates (>1 petagrams [Pg] of carbon per year; 1 Pg = $10^{15}$ grams).

Furthermore, with the rapid rate of Arctic warming and reduced seasonal and permanent sea ice today, the modern ocean may be approaching a tipping point in which the sinking of water in the North Atlantic, an important driver of the Atlantic meridional ocean circulation (AMOC; see Figure 1.2 in Chapter 1), may cease. Present model forecasts offer different perspectives on potential future changes (Ditlevsen and Ditlevsen, 2023). Forecast differences are perhaps due to limited direct observations (F. Li et al., 2021) and minimal understanding of tipping points in global ocean circulation and their broader consequences.

**Given the importance of the ocean to meridional heat transport** (Trenberth and Caron, 2001) **and carbon cycling** (Sigman and Boyle, 2000; Toggweiler, 1999), **climate and Earth system models that investigate these interrelated processes, and changes that may occur, require validation based on known past scenarios, knowledge that can be gained only by scientific ocean drilling. This is perhaps the most urgent goal of ocean drilling today.**

*Progress Made During IODP-2*

Several expeditions have provided key contributions to understanding past climate and ocean change, with implications for understanding future ocean change. Collectively, these results have fundamentally improved understanding of the linkages between the ocean, climate, and rates of climate change relevant to the challenges facing society today and in the near future. Such work was possible in large part because of research involving paleoclimate proxies of climate and ocean variables. The proxy data collected and measured essentially serve as surrogates, or indirect indicators, of past changes in temperature, ice volume, and ocean chemistry, among others.

Expeditions to the Southern Ocean directly addressed the past dynamics of ice sheets and sea levels. These included expeditions to the Ross, Amundsen, and Weddell seas, which yielded several key advances. Cores from the Ross Sea were used to reveal the previous presence of a large West Antarctic Ice Sheet (WAIS) during the middle Miocene (18–14 Ma; Figure 3.6), which could account for the 40- to 60-meter sea level variations previously documented from both pelagic and continental margin records. In the absence of a WAIS, such large-amplitude sea level changes would require complete loss of the East Antarctic Ice Sheet, something that could not be achieved in models, even under high levels of $CO_2$ (Marschalek et al., 2021). Furthermore, analysis of cores from the Amundsen Sea revealed significant retreat of the WAIS during the middle Pliocene warm period, 3.3–3.0 Ma (Gohl et al., 2021).

Expeditions designed to assess the sensitivity of the climate system to higher greenhouse gas levels in the past (i.e., 60–3 Ma) included coring of the western equatorial Pacific and South Pacific, and along the South Africa margin. Analysis of the sediment cores collected in the Pacific addressed several questions related to the role and response of the western Pacific warm pool (i.e., especially warm surface waters in the western equatorial Pacific Ocean) to variations in greenhouse gases, on millennial and longer timescales. Reconstructions showed that regional sea surface temperature varied in sync with greenhouse gas levels over the last 12 myr, including during the Holocene. This finding helped resolve a critical discrepancy between models and previous reconstructions of Holocene global temperatures. Cores recovered along the South Africa margin supported models that attributed redistribution of salinity differences between the ocean basins as a primary factor in driving changes in global ocean circulation patterns during glacial periods.

Changes in precipitation patterns (i.e., hydroclimates) associated with climate change will have profound impacts on society, particularly at the regional scale. Such systems include seasonal monsoons and atmospheric

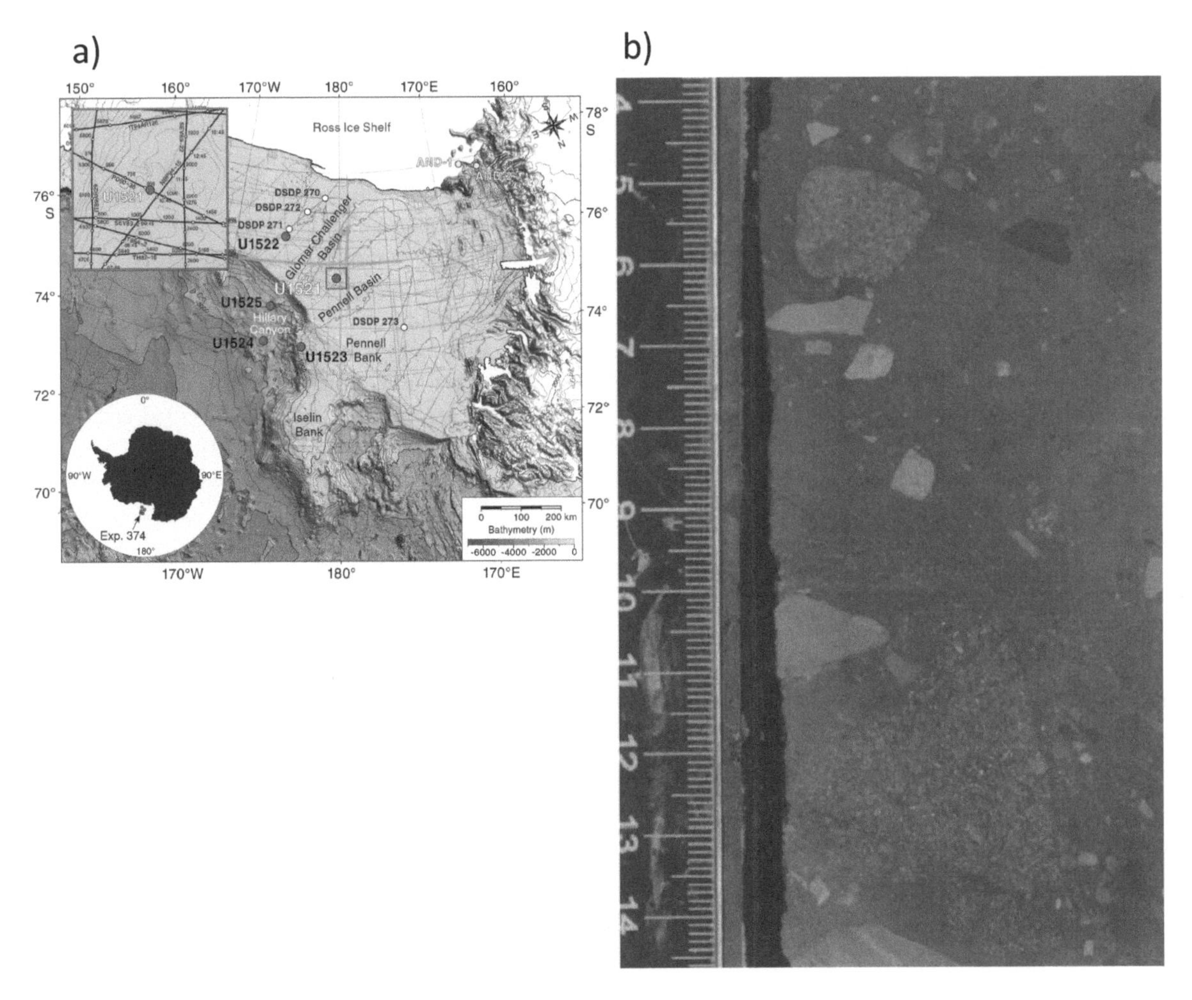


**FIGURE 3.6** (a) Map of sites cored in the Ross Sea during Expedition 374. (b) Photo of core containing Miocene clast-rich sandy diamictite taken from Site 1521 (Unit VII 63R-CC, 4–14 cm). NOTES: The diamictite is an ice-proximal glaciomarine deposit; the clasts are eroded and transported pebbles that rained out from floating ice or from subglacial deposition. The diamictite is evidence for a larger West Antarctic Ice Sheet in the Miocene—just before $CO_2$ levels increased.
SOURCE: McKay et al., 2019.

rivers that impact billions of people. Informing models on the evolution of regional monsoons over glacial–interglacial cycles in response to the warming caused by the increase of greenhouse gases was the goal of expeditions to the Indian Ocean, Arabian Sea, and Maldives. Major scientific contributions from these expeditions include establishing the timing and origin of the onset of the modern South Asian monsoon. This work provided new understanding of the evolution of Plio-Pleistocene summer monsoon rainfall, leading to better model-predicted increases in monsoon precipitation and variability due to greenhouse gas forcing. Furthermore, data from these cores were used to demonstrate how high-latitude cooling around Antarctica from 12 to 8 Ma initiated major changes in precipitation patterns in Australia and Southeast Asia.

Despite an increase in the atmosphere's vapor-holding capacity with increasing temperature, the hydrologic cycle is expected to amplify meridional vapor transport and precipitation cycles while shifting major rain patterns (Held and Soden, 2006; Trenberth, 2011). Although all models show hydrologic intensification, the exact patterns and magnitudes of change vary from model to model. Verification of the sensitivity of the hydroclimate to global warming has come mainly from paleo observations of the recent and more distant past. The paleo observations are largely preserved in ocean basins where wind and runoff deposit the signals of the local hydroclimate. Evidence for larger-scale changes in the hydrologic cycle are best preserved in deep-sea archives, particularly during extreme warm periods. Model simulations of these extremes show that increased meridional vapor transport resulted in significantly steeper meridional sea surface salinity gradients, with higher salinities in the tropics and lower salinity at high latitudes (Carmichael et al., 2016, 2017), not unlike forecasts for the future.

To address additional model uncertainties, the international climate modeling community initiated a coordinated effort to compare all major models using select reconstructions of past climates. This effort, designated the Paleoclimate Model Intercomparison Project (PMIP), was established to evaluate the models; understand the model–model and model–data differences; and, where possible, provide suggestions for model improvements. Following protocol, PMIP focused initially on the last glacial–interglacial transition, but it eventually extended its work to focus on the extreme greenhouse intervals (e.g., middle Miocene, Eocene, and Paleocene–Eocene thermal maximum [PETM]), periods when $CO_2$ levels were in the range expected by the year 2100 (600–1,000 ppm; Figure 3.5). This effort, called the Deep-Time Model Intercomparison Project, has contributed to a better understanding of the mechanisms of climate change and the role of climate feedbacks (Hollis et al., 2019; Lunt et al., 2017). It has provided the first robust paleo-based estimates of Earth climate sensitivity (Lunt et al., 2017; Zhu et al., 2021). The most recent studies utilizing reconstructions of sea surface temperatures and $CO_2$ are derived largely from sediment samples recovered by scientific ocean drilling (Hollis et al., 2019) and suggest an Earth climate sensitivity closer to the high end of 3.5–4.0°C per doubling of $CO_2$.

Finally, observations of the last several hundred thousand years collected by scientific ocean drilling reveal millennial-scale ocean circulation changes with far-reaching impacts. For example, changes in heat transport potentially impact ice sheets in the opposing hemisphere, as well as global atmospheric circulation and hydroclimate in general (e.g., Brahim et al., 2022).

*High-Priority Future Research*

**Additional proxy-based observations obtainable only by scientific ocean drilling are required to assess the accuracy of climate models in replicating greenhouse gas–forced changes, including tipping points in ice sheet dynamics and sea level change, ocean circulation, and global temperatures** (Figure 3.5; Box 3.2). In fact, the tipping points in Earth's interconnected climate system require greater understanding (McKay et al., 2022). Responses to climate forcings can be nonlinear; what may be gradual change initially can shift to rapid change if a critical threshold, or tipping point, is reached. Furthermore, observational gaps remain for the past extreme greenhouse gas periods in two climatically sensitive regions: the Arctic, where long cores that span the necessary time periods are from a single subseafloor site, and the equatorial ocean, where a single record from the Indian Ocean suggests that coastal ocean sea surface temperatures might have exceeded 40°C during the PETM.

Previous reconstructions of ocean circulation of the last glacial maximum established the presence of stable modes of the AMOC with weaker deep-water production in the North Atlantic (Böhm et al., 2015). However, the climatic conditions (e.g., temperature, sea surface salinity) that define the bounds of tipping points (i.e., mode switches) for the hydrographic parameters in areas of deep-water formation remain poorly understood. Such an abrupt rearrangement of large-scale circulation has the potential to impact climate regionally and globally; thus, understanding these parameters is important. In addition, structural changes in the deep circulation may significantly impact the accumulation and return of nutrients and oxygen and carbon dioxide to the surface ocean and thus influence marine biological productivity and carbon fluxes. **A greater understanding of the past major circulation regimes would provide insight into the potential modes of circulation possible in the future**. For example, recent simulations of the early Eocene circulation using an advanced Earth system model (Zhang et al., 2020) show enhanced ocean heat transport and Southern Ocean warming due to a more vigorous rate of overturning compared with simulations using other models (e.g., Winguth et al., 2012).

Accurate predictions of how much global mean sea level (GMSL) could rise in the near future are not possible based only on modern observations. While thermal expansion of seawater and melting of glaciers have dominated

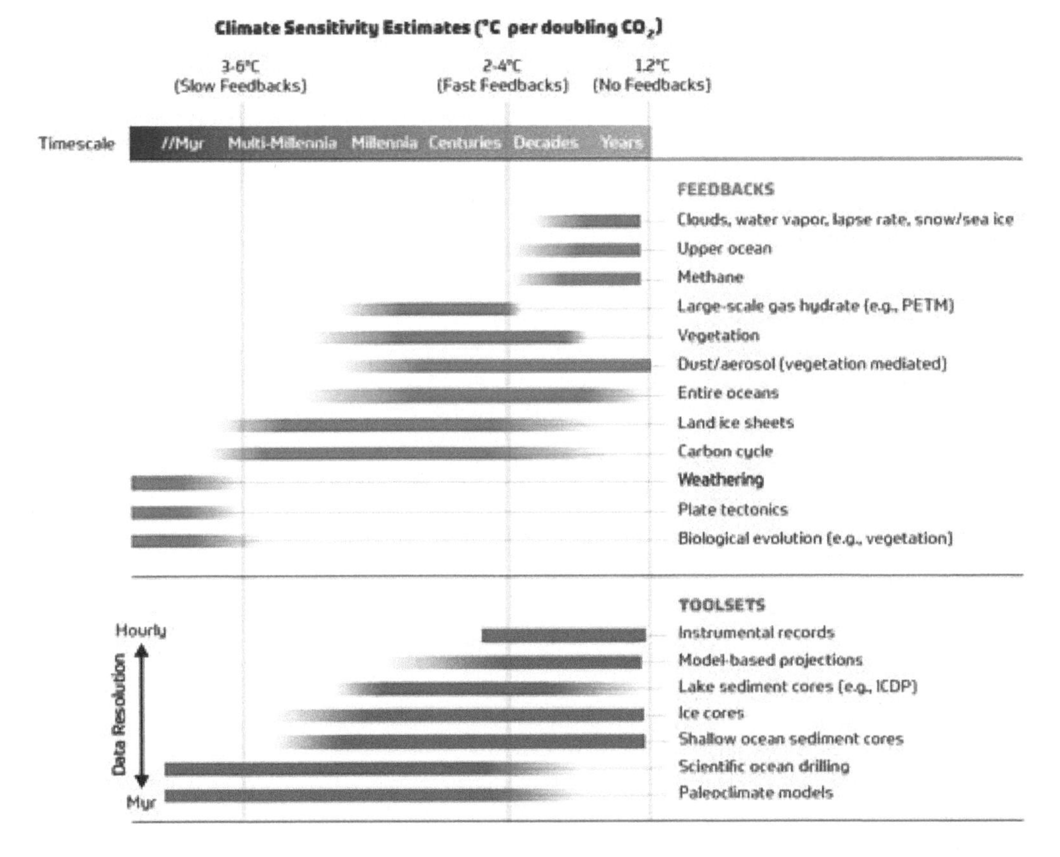


**FIGURE 3.7** Feedbacks and toolsets for estimating climate sensitivity. NOTES: Climate sensitivity is one of the most important quantitative estimates of climate change. Climate system feedbacks (blue bars) operate over timescales of years to many thousands of years (and longer). Different toolsets (purple bars) are therefore needed to investigate the feedbacks. Compared with most other toolsets, scientific ocean drilling allows investigation of the influence of Earth system feedbacks on climate sensitivity over a wider range of timescales—including those for which equilibrium climate states of extreme warmth existed. ICDP = International Continental Scientific Drilling Program; myr = million years; PETM = Paleocene–Eocene thermal maximum.
SOURCE: Modified from Rohling et al., 2012.

GMSL rise over the last century, mass loss from the Antarctic and Greenland ice sheets is expected to exceed other contributions to GMSL rise under future warming scenarios (Dutton et al., 2015). Therefore, forecasting the response of these ice sheets and GMSL to warming remains an important, yet challenging, task. The challenges stem partly from an incomplete understanding of ice sheet and ice stream internal dynamics, as well as the role of heating from above and below. Moreover, recent advances in representing the dynamics of ice loss at shelf edges suggest a much higher sensitivity in sea level response to small changes in temperature at the regional scale (DeConto and Pollard, 2016; DeConto et al., 2021). As such, a better understanding of how the lost mass of ice sheets contributed to sea level rise during past warm periods can constrain the process-based models used to project ice sheet response and sensitivity to future climate change (Figure 3.8). **Scientific ocean drilling is the only means of reconstructing ice sheet changes during analogous times of warming in the deeper past.** While efforts thus far have focused largely on recent interglacials (with emphasis on 200,000–100,000 years ago), when sea level was higher despite similar levels of preindustrial $CO_2$ (280–300 ppm), more recent studies have focused on times when ice sheets existed, and $CO_2$ levels were in the range projected for the present or near future

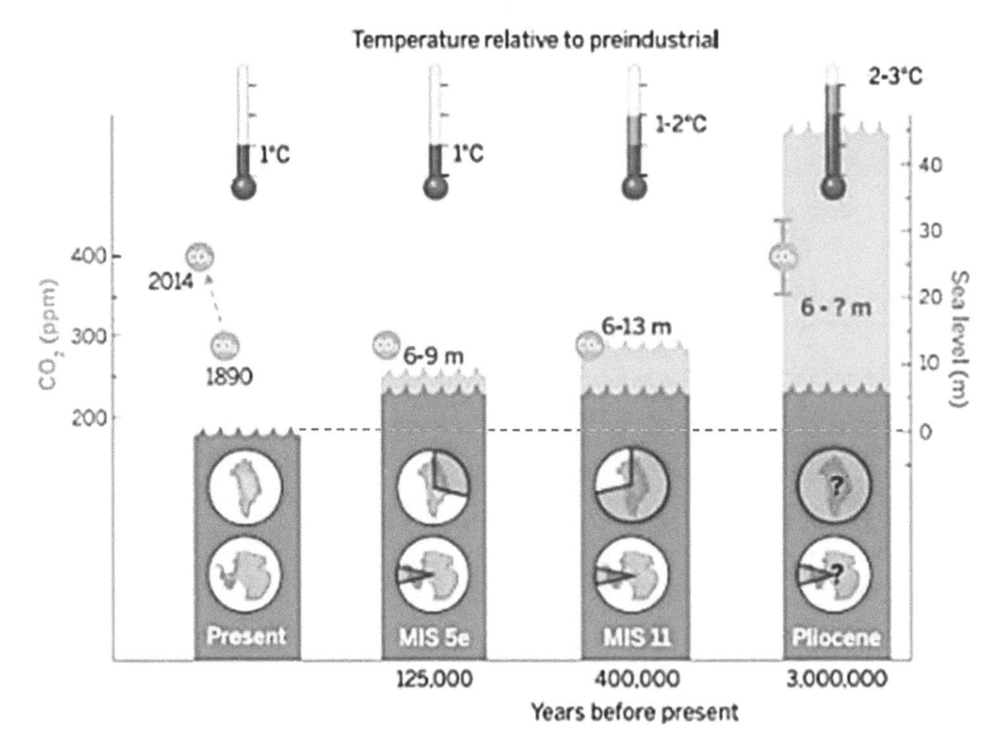


**FIGURE 3.8** Peak global mean temperature relative to preindustrial, atmospheric $CO_2$ (preindustrial = 280 ppm, 2014 = 397 ppm, 2023 = 420 ppm), maximum global mean sea level (GMSL, with present day = 0 m) and source(s) of meltwater from Greenland and Antarctica. NOTES: Light-blue shading indicates uncertainty of GMSL maximum. Red pie charts over Greenland and Antarctica denote fraction (not location) of ice melt, contributing to sea level rise. MIS = marine isotope stage. SOURCE: Dutton et al., 2015.

(400–800 ppm; Figure 3.5). Crucially, the latest reconstructions of the Antarctic landmass extent and elevation during these warm periods have been key to reconciling the changing sensitivity to greenhouse gas forcing with the present (Marschalek et al., 2021). Incorporating constraints on ice sheet extent and estimates of regional sea level to constrain and test models of these warm periods will require future scientific ocean drilling.

**Reconstructing sea surface salinity gradients during periods of elevated greenhouse warming would be a key test of first-order predictions of how hydroclimates change under greenhouse climate states.** A minor misrepresentation of the dynamics of vapor transport could seriously bias climate models in multiple ways. In this regard, efforts to establish past changes in vapor transport would benefit from more detailed spatial (meridional) sediment cores spanning the geography of net evaporation and precipitation during the extremes. In yet another observations gap, core records of monsoon systems have been obtained mainly from the Northern Hemisphere (primarily from ocean basins in South Asia). **Collection must be expanded to the Southern Hemisphere, and the cores must be deep enough to sample further back in time to periods of extreme warmth in both hemispheres.**

**CONCLUSION 3.3a** Additional observations obtainable only by scientific ocean drilling are required to assess the skill of climate models to replicate greenhouse gas–forced switches (i.e., tipping points) over geological timescales in temperatures, ice sheet dynamics, sea level, and ocean circulation and to constrain the role of feedbacks (physical or biogeochemical responses that amplify or dampen perturbations). To constrain Earth climate sensitivity to high greenhouse gas levels, additional scientific drilling is required to fill data gaps for extreme warm intervals in climatically sensitive regions (e.g., the Arctic and equatorial oceans, and in a few cases, the midlatitudes). Similarly, to fully characterize the sensitivity of hydroclimates (including regional monsoons) to greenhouse gas forcing, records obtained for the Northern Hemisphere need to be complemented with records from the Southern Hemisphere.

### Evaluating Past Marine Ecosystem Responses to Climate and Ocean Change

*Using fossils to determine ecosystem responses to past environmental drivers (warming, ocean acidification, and deoxygenation)—a lens informing the future.*

Seafloor sediment microfossils (i.e., small, mineralized fossils) and molecular fossils preserve a history of marine biodiversity, including the origin and extinction of species. They are used to better understand how climate and ocean changes affect the evolution of life and ecosystems, and of marine biodiversity and distributions through long periods of time. Determining the timing of extinction and speciation events through microfossils is essential for further developing regional age-depth models (allowing paleoceanographers to convert subseafloor depth to age and determine sedimentation rates), as well as to tracking ecosystem evolution and reconstructing past ocean conditions.

The future impacts of rapid global change on marine ecosystems are unknown, but some insights can be gained from studying past long-term environmental perturbations, which have influenced the evolution of marine organisms and ecosystems. Understanding the details of ecosystem response to these past events also provides opportunities to test advanced Earth system and ecosystem models designed to simulate the impacts of continued anthropogenic carbon emissions and global warming on marine ecosystems. Indeed, the closest analogs to Earth's likely future are the transient climate events known as hyperthermals, lasting 10,000 –20,000 years, that occurred during the early Cenozoic interval of elevated warmth (~60–40 Ma) (Norris et al., 2013).

Looking forward, the combination of warming-induced changes in ocean circulation and stratification coupled with acidification and deoxygenation has the potential to significantly modify ocean ecosystem structures across the planet. The most severe impacts might result from cascading effects on large-scale biogeochemical processes—such as microbial respiration, particle remineralization and $O_2$ consumption, denitrification and nitrogen fixation—and hence, biological export production (Henson et al., 2022; Hutchins and Capone, 2022; Thomalla et al., 2023). These changes would be in addition to the more predictable first-order poleward shifts in biogeographic ranges of most species. Furthermore, the ecological impacts will be significantly compounded at the coasts by changes in regional precipitation, runoff, and nutrient supply. All these changes have the potential to constrict habitability for most marine taxa.

**Past perspectives derived from scientific ocean drilling improve understanding of the potential range of ocean states, providing insights into the historical range of ecosystem responses to changes in ocean and climate conditions** (Halpern et al., 2015)**, and informing ecosystem models**. Dozens of metrics, indicators, and even thresholds delineate ocean ecosystem conditions in modern times (Rice and Rochet, 2005), including the presence of common chemicals, abundance of key biota, rates of key ecological processes, and emergent properties of marine ecosystems (e.g., biodiversity); but it is only from past records of ocean conditions analogous to those predicted that paleobiologists can gain direct evidence of what the future may hold for marine ecosystems.

Global warming will likely lead to vertical compression of upper-ocean ecosystems via a weakened biological pump[3] (Figure 3.9). Under normal conditions, with a thermally stratified water column, a substantial fraction of

[3] The *biological pump* is a set of processes by which the ocean biologically sequesters atmospheric carbon from surface waters into the ocean's interior.

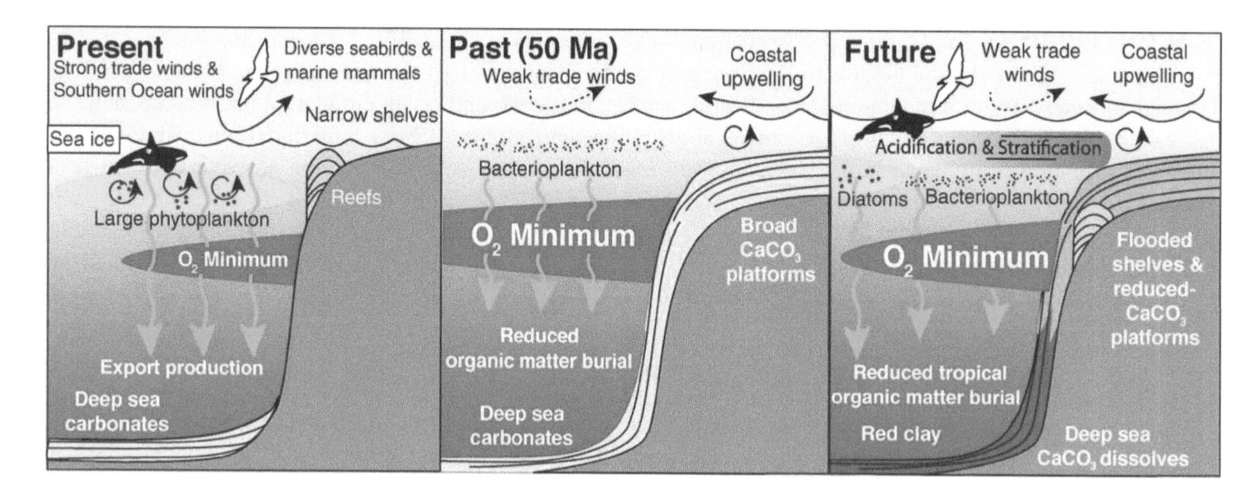


**FIGURE 3.9** Comparison of present, past, and future ocean ecosystem states. NOTES: In the geological past (middle panel), a warmer, less oxygenated ocean supported longer food chains (or food webs) based in picophytoplankton—primary producers smaller than present-day phytoplankton (left panel). The relatively low energy transfer between trophic levels in the past made it hard to support diverse and abundant top predators dominated by marine mammals and seabirds and reduced deep-sea organic matter burial. Equilibration of weathering with high atmospheric $CO_2$ allowed carbonates to accumulate in parts of the deep sea. Reef construction was limited by high temperatures and coastal runoff even as high sea level created wide, shallow coastal oceans. In the future (right panel), warming will eventually reproduce many features of the past warm world but will also add transient impacts such as acidification and stratification of the surface ocean. Acidification will eventually be buffered by dissolving carbonates in the deep ocean, which create carbonate-poor "red clay." Stratification and the disappearance of multiyear sea ice will gradually eliminate parts of the polar ecosystems that have evolved in the past 34 Ma and will restrict the abundance of large phytoplankton-based food webs that support marine vertebrates in the polar seas.
SOURCE: Figure and modified caption from Norris et al., 2013.

sinking particle fluxes escapes remineralization in the surface ocean mixed layer, enabling various plankton (and benthic) species to survive at depths well below the photic zone, or in what is commonly referred to as the twilight zone. This is also the depth at which the level of dissolved oxygen is reduced via respiration, creating an oxygen minimum zone (OMZ) and oxygen deficient zones (ODZs). In theory, with warming of the surface ocean, nutrient delivery from upwelling will decrease and the rate of remineralization of organic detritus in the upper ocean will increase, thus reducing the sinking particle flux and forcing deep-dwelling plankton closer to the surface (vertical compression). In addition to the impact on species distribution and ecosystem structure, a weakened biological pump will reduce the extraction of $CO_2$ from the surface ocean, a potential positive (i.e., reinforcing) feedback on global warming. This state, with decreased consumption of oxygen by respiration in the dark ocean, could potentially represent the equilibrium state for a warmer ocean, at least as simulated by Earth system models, whereas the nonequilibrium transient state might be characterized by deoxygenation due to enhanced stratification.

*Progress Made During IODP-2*

Microfossil studies during IODP-2 that utilized newly recovered marine sediment archives and those already stored in core repositories documented environmental changes and their ecosystem responses on a range of timescales and oceanic settings. Perhaps more dramatic, scientific ocean drilling has uniquely documented extinction and the rapid recovery of marine benthic and planktic life at the Chicxulub impact crater (Lowery et al., 2018) (Box 3.3). Other work on microfossils from globally distributed sites has demonstrated that the efficacy of the biological pump and carbon cycling in the upper ocean was strongly controlled by temperature in the late Neogene (past 15 myr) (Boscolo-Galazzo et al., 2022). Globally distributed marine microfossil records have also revealed

## BOX 3.3
## Drilling the Cretaceous–Paleogene Impact Crater and Documenting the Demise and the Recovery of Life

The dinosaurs were not the only forms of life that went extinct 66 million years ago when a 10-km asteroid struck what is now the Yucatan Peninsula and continental shelf waters; 76 percent of all species on Earth were eradicated during the end-Cretaceous mass extinction. Dramatic details of the widespread catastrophic event were documented pristinely during Ocean Drilling Program Expedition 171 from sediment cores in the western North Atlantic Ocean—at a site that was ~1,500 km away from point of impact (Norris et al., 1998). A replica of the famous Expedition 171 core section is on display in the Smithsonian Museum of Natural History Ocean Hall. Eighteen years later, another major accomplishment took place: International Ocean Discovery Program Expedition 364 (Morgan et al., 2016) used a unique mission-specific "lift boat" to drill into the exceptionally preserved Chicxulub impact crater. The cores recovered from this site are astonishing. Expedition drilling evidence shows that when the Chicxulub asteroid hit, the Earth rebounded, bringing up pink granite (Figure 3.10, left core photo) from 10 km below the surface, which collapsed around the center of the crater to form concentric rings. Computed tomography scans showed that the fractured and porous rock had many pathways for fluids, making it an intriguing place to look for the recovery of life, in the form of microbes in the peak ring. Paleontological and geochemical studies of cores several hundred meters above these basement impact structures documented how this large impact affected ecosystems and biodiversity at ground zero. Microfossils in the impact site sediment layers (Figure 3.10, right core photo) provided strong evidence for rapid recovery of life at ground zero: benthic and planktonic life reappeared in the basin just years after the impact, and a high-productivity ecosystem was established within 30,000 years (Lowery et al., 2018). Such findings demonstrated that proximity to the impact was not a control on biological recovery. Instead, natural ecological processes probably controlled the recovery of productivity after the Cretaceous–Paleogene mass extinction and are therefore likely to be important for the response of the ocean ecosystem to other rapid extinction events, such as climate-related changes in ocean chemistry that are impacting modern ocean health.

*continued*

## BOX 3.3 Continued

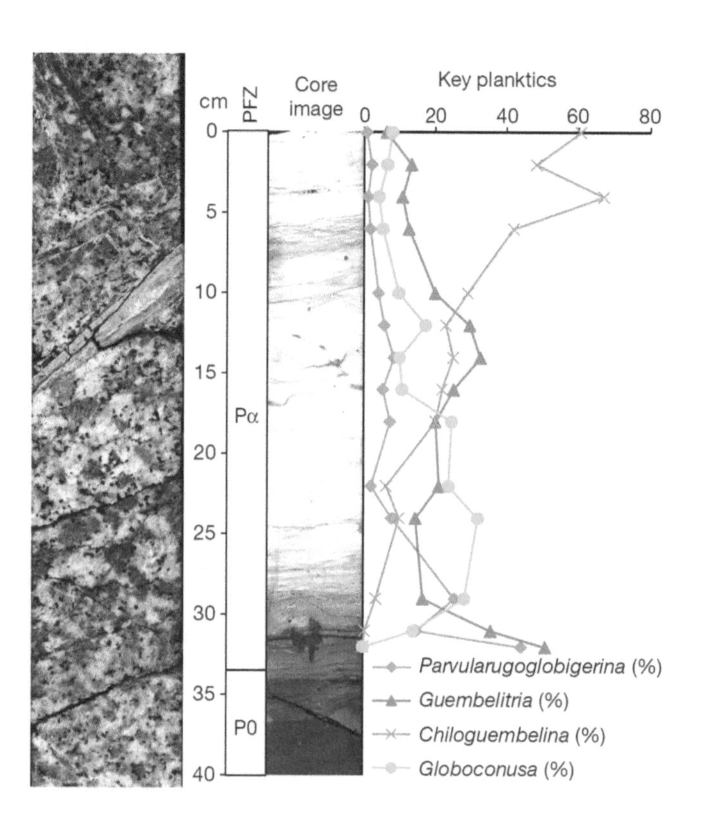


**FIGURE 3.10** (Left) Pink granite brought up from 10 km below the surface when the asteroid hit the Yucatan Peninsula and continental shelf waters.
SOURCE: University of Texas Jackson School of Geosciences. (Right) Paleontological evidence of the recovery of planktic life at the impact site during the earliest Paleogene. Percent abundance of key planktic foraminifera groups are shown. Darker rock is a transitional unit, and the white rock is a pelagic limestone. P0 and Pα are tropical planktic foraminiferal biostratigraphic zones (PFZ) (Wade et al., 2011). Because many planktic foraminifera species originate at or near the base of Pα, it was concluded that the base of the limestone lies very near the base of this zone.
SOURCE: Lowery et al., 2018. Reproduced with permission from Springer Nature.

that plankton evolution and diversity have been paced by orbitally forced changes in climate and the carbon cycle over the last several million years (Woodhouse et al., 2023). Recent work on cores from the Southern Ocean has identified antiphased dust deposition and biological productivity over 1.5 myr (Weber et al., 2022).

Biogeochemical cycling can be tracked through time with sediment samples and data procured from scientific ocean drilling, as microfossils can indicate how past conditions impacted ocean conditions. For instance, investigations of changes in plankton community structure between warm and cold periods over the last 66 myr reveals that plankton were less abundant and diverse and lived much closer to the surface during warm periods (Crichton et al., 2023) (Figure 3.11). In contrast, during the long-term cooling trend of the Cenozoic, fossil evidence indicates increased plankton species diversity and greater export of detrital organic carbon to the deep ocean. These results are consistent with models on the vertical compression of habitats as the ocean warms and their expansion as the ocean cools (e.g., Crichton et al., 2023). This paleo-reconstructed relationship is significant because the biodiversity of plankton in the modern ocean is correlated to tuna, billfish, krill, squid, and other key fauna (Yasuhara et al., 2017), suggesting food web implications as the modern ocean continues to warm.

Planktic foraminifera are a major constituent of ocean floor sediments and thus provide one of the most complete fossil records of any organism. IODP-2 expeditions to sample these sediments have produced large amounts of spatiotemporal occurrence records throughout the Cenozoic. **These scientific ocean drilling program data are the primary source of the newly established Triton database, which has been populated with more than 50,000 records of species-level occurrences of planktic foraminifera**. The database can now be used to study how species responded to past climatic changes.

Relevant to the question of deoxygenation, over several decades **scientific ocean drilling has recovered evidence of expansive ocean anoxia during the Cretaceous** (80–120 Ma) when layers of the upper ocean lacked sufficient oxygen to support respiration or to remineralize detrital organic matter, which accumulated in thick layers known as black shales. More recently, biomarkers (molecular fossils) extracted from these shales demonstrate how organic carbon burial drivers, such as enhanced productivity and/or preservation, operated along a continuum in concert with microbial ecological changes. Localized increases in primary production can trigger marine microbial reorganization from the surface waters to the seafloor, and can destabilize carbon cycling, promoting progressive marine deoxygenation and ocean anoxia events (Connock et al., 2022). These Cretaceous ocean anoxia events lasted for hundreds of thousands of years and were likely triggered by excessive emissions of $CO_2$ and nutrients (e.g., iron) associated with volcanism and massive extrusion of basalts (Figure 3.12). In contrast, during other warm periods, such as the Eocene or PETM for example, the OMZ appears to have been relatively well oxygenated with less denitrification (Kast et al., 2019). The reason for the contrasting conditions remains enigmatic but likely involves a combination of factors, including differences in the ocean circulation and mixing and overall nutrient inventory (Auderset et al., 2022).

Additionally, microfossil records documenting past ocean ecosystems and marine communities have been used to reconstruct ancient ocean structures and circulation in a range of settings and timescales, thus demonstrating how research on paleobiological components and research on paleoclimatic and paleoceanographic systems are interdependent. For example, during IODP-2, a 1.5-million-year-old northern Indian Ocean drilling core record, containing summer and winter planktic foraminifera microfossil assemblages, documented a climatically triggered large-scale reorganization of the Indian monsoon system at the time of the middle Pleistocene transition (i.e., the time when glacial–interglacial cycles became both more extreme and paced with a longer periodicity; Bhadra and Saraswat, 2022). On a deeper timescale, foraminifera depth habitat ecologies in an IODP-2 southern Indian Ocean drilling core record documented differential effects of a late Cretaceous $CO_2$-driven global cooling transition on surface versus deeper water in the southern high latitudes, consistent with enhanced meridional circulation (Petrizzo et al., 2022). A final example draws on planktic foraminiferal assemblage data collected from legacy cores during IODP-2. Data from sites that transect the modern-day Kuroshio Current and Extension were used to reconstruct diversity curves within the regional western boundary current through the last 12 myr. Results point to potential causal links between diversity gradients and variations in the regional western boundary current associated with tectonically driven ocean gateway closure and paleoclimatic events affecting ocean thermal gradients (Lam and Leckie, 2020). **Because "multidecadal variability of the strength and position of western boundary currents and short records from direct observations obscure the detection of any long-term trends"** (Gulev et al., 2021,

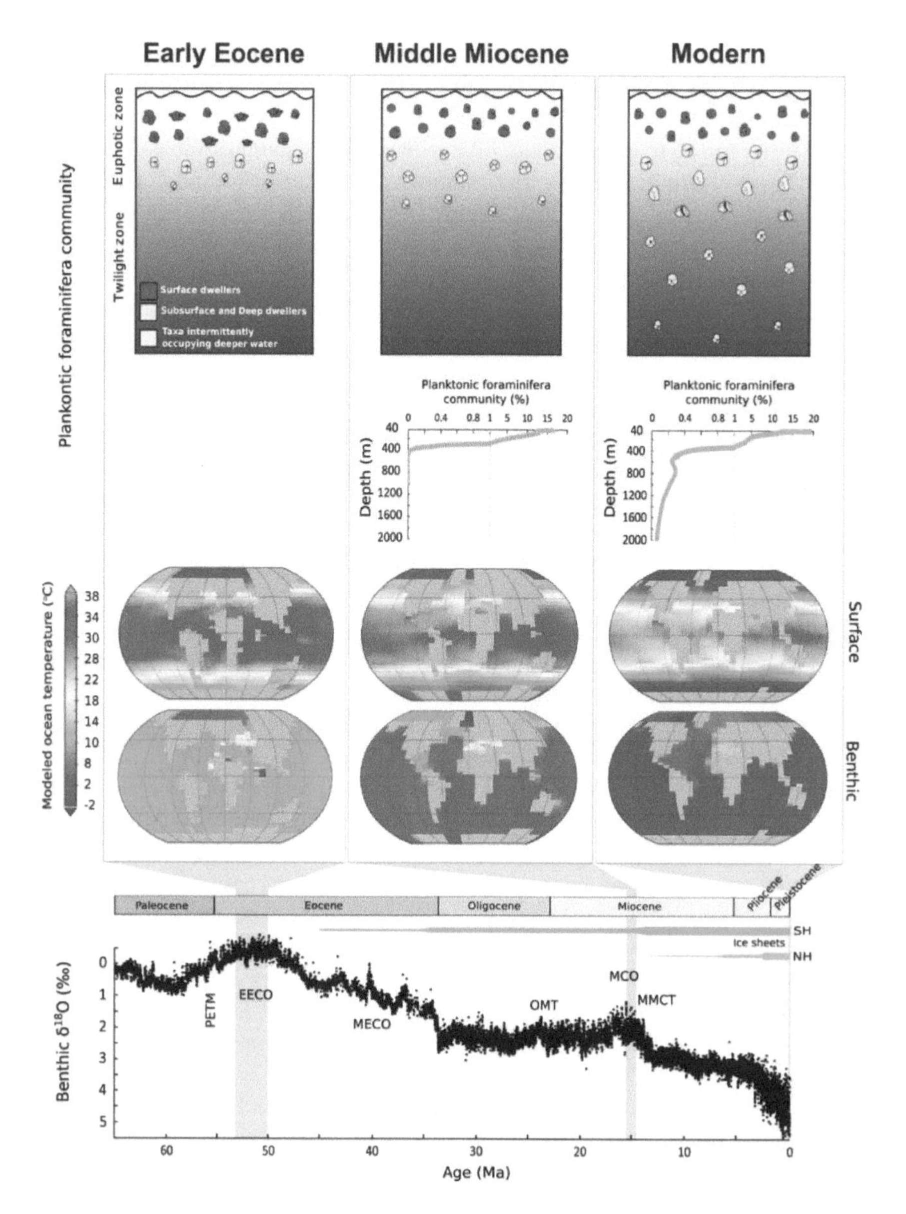


**FIGURE 3.11** Illustrations of planktic foraminiferal distribution and species diversity across the Cenozoic (last 66 million years). (Top panel) Conceptual diagram representing planktonic foraminiferal community distribution in the euphotic (0–200 m) and twilight (200–1,000 m) depth zones for the early Eocene, middle Miocene, and modern. (Second panel) Percent (%) abundance for the middle Miocene (pink) and modern (core top; blue). (Third panel) Modeled surface and bottom-water (benthic) ocean temperature and continental configurations for early Eocene, middle Miocene, and preindustrial present ("modern"). (Lowest panel) Global benthic $\delta^{18}O$ record, which is a proxy for bottom-water temperatures and global ice volume; higher values of $\delta^{18}O$ are interpreted as colder temperatures and greater ice volume (when planet is not ice free). NOTES: EECO = early Eocene climatic optimum; Ma = million years ago; MECO = middle Eocene climatic optimum; MCO = Miocene climatic optimum; MMCT = middle Miocene climate transition, as recorded in deep ocean sediment records; NH = Northern Hemisphere; OMT = Oligocene–Miocene transition; PETM = Paleocene–Eocene thermal maximum; SH = Southern Hemisphere.
SOURCE: Crichton et al., 2023.

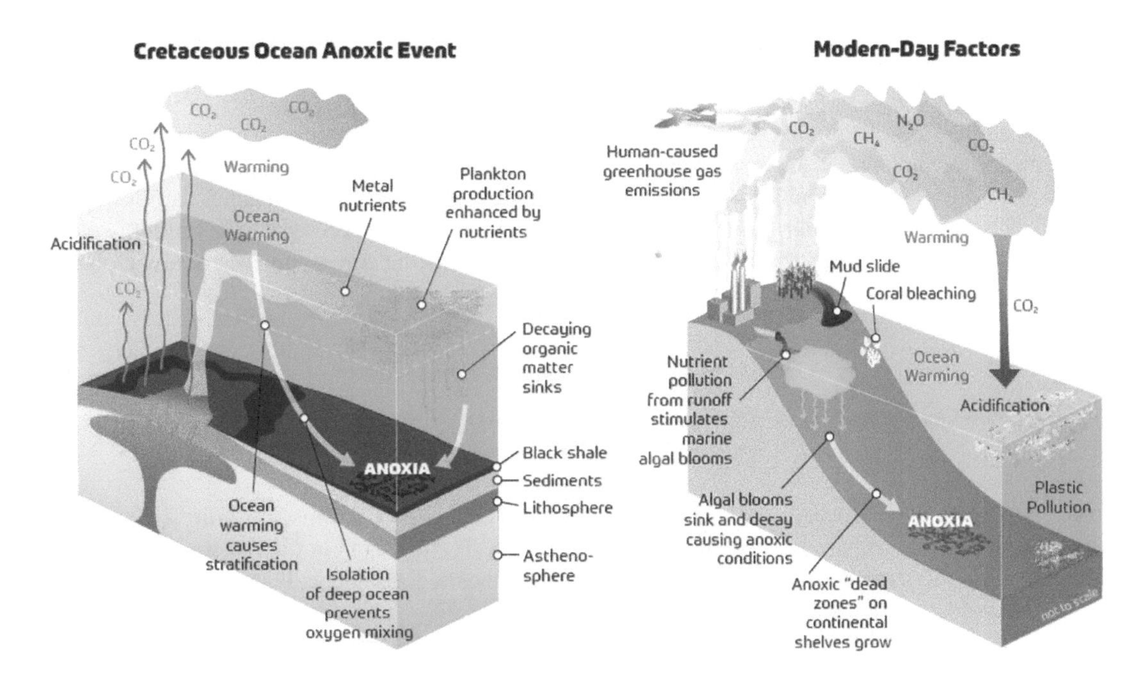


**FIGURE 3.12** Schematic comparison of past conditions that resulted in ocean anoxic events (left), and modern-day environmental perturbations (right). NOTES: While the source of the $CO_2$ was different during the Cretaceous (massive volcanism associated with the emplacement of large igneous provinces in the Cretaceous vs. anthropogenic-sourced $CO_2$ today), the ocean system response of the past could be predictive of the future.
SOURCE: Koppers and Coggon, 2020. Illustration by Rosalind Coggon and Geo Prose.

p. 357), **scientific ocean drilling paleobiological work helps fill a gap in research on the marine ecosystem responses to climate and ocean change**.

*High-Priority Future Research*

Ocean drilling has provided key insights into marine ecosystem responses to changes in past ocean and climate conditions, with potential implications for the future of ocean health in a rapidly warming world. However, a number of issues regarding the impacts of environmental extremes on past ocean ecology remain unresolved. Some can potentially be addressed with existing archives, but others require new data. **One issue in particular is the question of habitability of the tropics**, and threshold temperatures for phyto- and zooplankton. To address this, **additional drilling is required to target the short-lived extremes in equatorial oceans**, particularly along the continental margins, where (e.g., during the PETM) various groups of plankton (foraminifera and dinoflagellates) appear to have abandoned the surface ocean, reappearing only after temperatures cooled. **Identifying an upper thermal limit or range for habitability of the tropical ocean during the past remains a high-priority challenge**.

Similarly, there is still a need for additional examples in the paleo record to be uncovered that display how past plankton communities shifted poleward during times of past warming. The recent geographic ranges of marine organisms, including planktic foraminifera, diatoms, dinoflagellates, copepods, and fish, have been seen to shift poleward because of climate change. However, it remains unclear the extent to which the poleward move represents precursor signals that may lead to extinction. Additionally, some of these shifts are taking place in midlatitude ecotones, places where warm subtropical and cool subpolar waters meet, areas that also have some of the highest biodiversity in the world today (Tittensor et al., 2010). Therefore, subtropical to subpolar paleocommunities overlap in need of investigation. **A more in-depth understanding of the development of marine biodiversity patterns over time and space, and the influencing factors, is needed** (Woodhouse et al., 2023).

**CONCLUSION 3.3b** Additional scientific ocean drilling that prioritizes locations with limited records, such as the equatorial, midlatitude, and polar oceans and open ocean environments during past periods of extreme warmth, will allow paleobiologists to inform models of plankton ecosystem dynamics during past analog climate states (e.g., rapid warming). In addition, existing long-term paleo records can be further exploited for studies, capitalizing on the development of new databases and existing core samples to assess global marine ecosystem responses to climatic and oceanic shifts more fully.

## Monitoring and Assessing Geohazards

*Providing data to more accurately forecast and assess future risks of earthquakes, volcanic eruptions, submarine landslides, and tsunamis.*

Geohazards, including earthquakes, volcanic eruptions, landslides, and tsunamis, are a direct threat to human populations, with a record of harming people and damaging infrastructure, both in today's world and throughout human history. Thus, better understanding geohazards benefited society fundamentally by enabling more accurate and timely forecasting and assessment of future risks.

Tsunamis, landslides, earthquakes, and volcanic eruptions all create signatures in ocean sediments that can be sampled by ocean drilling. Through scientific ocean drilling, long-term instruments have been installed and deployed in the subseafloor, and fluids, sediment, and rock cores have been collected to study geohazards.

Among Earth's most hazardous tectonic environments are subduction zones, where one tectonic plate slides beneath another. These plate boundaries typically occur within or at the margin of ocean basins and are host to Earth's largest-magnitude earthquakes and most explosive volcanic eruptions. Seafloor motion associated with these events can also generate devastating tsunamis that impact both local and distant communities. Not only do these hazards pose significant danger to human life, but individual events also cause monetary damage, often exceeding $100 billion (e.g., the 2011 Mw [moment magnitude] 9.1 Tōhoku-oki [Japan] earthquake and subsequent tsunami killed 20,000 people and caused $210 billion in damages [Ranghieri and Ishiwatari, 2014]). Quantifying the risks associated with these hazards and developing better metrics for predicting when they may occur are of key societal importance as coastal populations continue to grow throughout the 21st century (Reimann et al., 2023).

Improving understanding of geohazards requires data from both the past and present. **Scientific ocean drilling can provide these data through (1) sedimentary records that provide constraints on the frequency and magnitude of past events, (2) direct sampling of rocks from fault zones or the slip plane of major landslides to determine the material properties of these features, and (3) real-time monitoring using borehole observatories.** These datasets complement other ongoing National Science Foundation (NSF) initiatives studying geohazards, including the Ocean Observatories Initiative's (OOI's) Regional Cabled Array, offshore Washington and Oregon; the Subduction Zones in Four Dimensions (SZ4D) initiative to study subduction systems in Chile, Cascadia, and Alaska; and several recently funded Centers for Innovation and Community Engagement in Solid Earth Geohazards (e.g., Cascadia Region Earthquake Science Center, Center for Land Surface Hazards, Collaborative Center for Landslide Geohazards).

Seafloor records are often more complete than onshore sediment sequences, as they are not exposed to subaerial erosion processes. Importantly, these signatures can be used to infer the magnitude of past events. For example, ground shaking generated by large earthquakes can cause slope failures that produce sediment turbidite deposits. Sampling the spatial distribution of these turbidites can provide an estimate of the magnitude of the earthquake. This approach has been used to constrain the recurrence interval of very-large-magnitude (M8 and M9) earthquakes off the coast of Cascadia (e.g., Goldfinger et al., 2012). Similarly, the magnitude of volcanic eruptions can be inferred from the thickness and distribution of ash records (Kennett et al., 1977). Deep cores recovered through ocean drilling can provide an important record of the size and frequency of past geohazards, allowing local communities to prepare for future events.

Direct sampling of fault zone rocks allows these rocks to be probed experimentally in the laboratory to determine their material properties. The IODP Nankai Trough Seismogenic Zone Experiment (NanTroSEIZE) used this approach to constrain the conditions that lead to a transition between stable fault creep at low strain rates to dynamic weakening at high strain rates. Results from the NanTroSEIZE project showed that once a rupture initiates, dynamic weakening mechanisms can drive rupture propagation to very shallow depths, thereby enhancing the likelihood of tsunami generation (e.g., Ujiie and Kimura, 2014). Future studies at other subduction systems will allow scientists to probe the different conditions (e.g., fault zone composition, fluid pressure) that promote seismogenic versus stable aseismic creep. These data can in turn be used to constrain numerical models of dynamic fault ruptures, earthquake cycles, and tsunami genesis.

Scientific ocean drilling also enables the installation of borehole observatories, which measure in situ subsurface conditions. Physical properties such as crustal stress and strain, temperature, pore pressure, and fluid chemistry have been observed to vary throughout a hazard cycle; in certain cases, they have been linked to precursory activity prior to a major event. For example, periods of "slow" or aseismic slip and/or enhanced subsurface fluid flow have been observed to occur before some large (≥ M8) earthquakes. These transients are difficult to resolve using instruments deployed on land, often far from the feature of interest, or directly on the seafloor where bottom currents generate noise that obscures the signal. By contrast, borehole observatories installed at depth, near fault zones or on the flanks of seafloor volcanoes, have significantly higher sensitivity; for example, **borehole sensors in the seafloor have an order of magnitude greater sensitivity than the pressure gauge sensors on the seafloor** (Box 3.4). While it remains unclear under what conditions precursory events occur, the ability to predict hazards requires future studies to better understand these signals in diverse geologic settings. Borehole observatories are thus critical to advance basic research into hazard cycles. Moreover, connecting observatories to cabled arrays can provide real-time monitoring of subsurface conditions with the prospect of developing future early warning systems. As such, borehole observatories complement OOI's Regional Cable Array off Cascadia and the aspirations of SZ4D to install similar infrastructure offshore of central Chile (Figure 3.13), which cannot be installed without scientific ocean drilling.

*Progress Made During IODP-2*

The IODP NanTroSEIZE project constrained the conditions that lead to a transition from stable fault creep at low strain rates to dynamic weakening at high strain rates (Box 3.5). Results from the NanTroSEIZE project showed that once a rupture initiates, dynamic weakening mechanisms can drive rupture propagation to very shallow depths, thereby enhancing the likelihood of tsunami generation (e.g., Ujiie and Kimura, 2014).

In the context of determining the controls on geologic hazards, a strategic objective of the IODP *2050 Science Framework* was to better understand the nature of slip processes. Faults can slip gradually or catastrophically or exhibit unstable (time and magnitude) behaviors. **In the last 10 years, major progress has been made in understanding slow-slip and tsunamigenic earthquakes, as a result of the NanTroSEIZE project and the installation of borehole observatories**, highlighted in Box 3.5. The NanTroSEIZE project, and related studies in the Nankai accretionary prism and shallow subduction interface, have taken place over multiple IODP expeditions.

Other expeditions to active subduction zones included one to the Sumatra subduction zone system, where an M9.2 tsunamigenic earthquake in 2004 destroyed many coastal communities. The properties of the materials being subducted at plate boundaries can determine where and when megathrust earthquakes occur and influence earthquake magnitude and tsunami hazard. Some of the most devastating recent earthquakes have not been found to conform to expectations with respect to their magnitude and tsunamigenic properties. For example, the thickness of the sediments that are being subducted between the Indo-Australian plate around Sumatra were not expected to result in large-magnitude earthquakes or tsunamis during a megathrust event. Yet, in 2004, the massive Sumatra-Andaman earthquake and tsunami killed 250,000 people. IODP-2 drilling in this area recovered two cores (extending down to 1,500 m below the seafloor) to characterize the sediment and rock properties of the material that is being subducted by the Indo-Australian plate. Geochemical analyses of the cores revealed that freshwater release from the dehydration of biogenic silica and silicate minerals during heating of subducting sediments may be influencing the strength of the fault. Such a process may have driven shallow slip offshore of Sumatra and

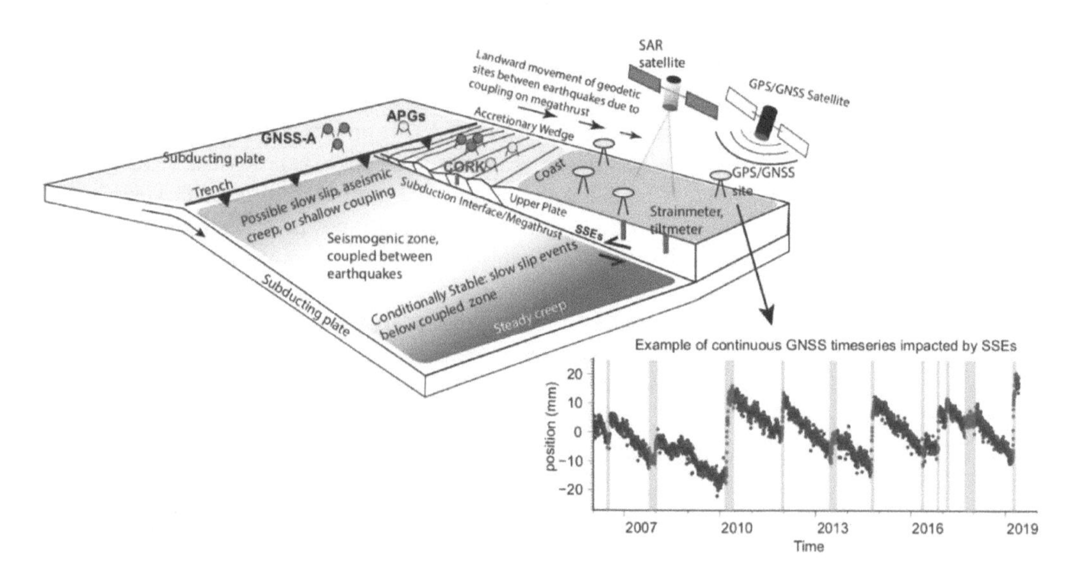


**FIGURE 3.13** Schematic diagram illustrating the types of geodetic sensors that can be used to distinguish between different slip behaviors on the subduction interface. NOTES: Bottom right inset shows an example of a continuous GNSS time series of subduction zone slip with slow-slip events highlighted by shaded blue bars. This diagram represents a composite of existing knowledge, but each margin has different aspects and questions that can be addressed by ocean drilling. APG = absolute pressure gauge; CORK = Circulation Obviation Retrofit Kit; GNSS = Global Navigation Satellite System; SAR = synthetic aperture radar.
SOURCE: Bartlow et al., 2021.

resulted in an increase in earthquake and tsunami size. These findings may be relevant for other subduction zones that exhibit similar sediment properties, such as Cascadia and the Eastern Aleutians.

Other **IODP-2 contributions have focused on mechanisms that control the occurrence of submarine mass failures (landslides), which can trigger tsunamis, destroy marine infrastructure, and alter carbon cycling in marine sediments**. Slope failure can be triggered by a variety of processes, including mechanical forcing (volcanic eruption, earthquakes, glacial-isostatic rebound), sea level change, rapid sedimentation, and fluid flow (e.g., gas hydrate dissociation). Uncovering the history of past regional submarine slides can aid in reconstructing the mechanisms of landslide initiation. Even submarine landslides around Antarctica could pose a tsunami risk to coastal communities in the Global South. It has been proposed that glacial–interglacial variations in sediment composition and sedimentation rates around the Antarctic continent could result in weak sediment layers that are more susceptible to landslides. The Iselin Bank, located in the eastern Ross Sea, sits on a passive margin (i.e., not near an active plate boundary). Seismic and chronological data obtained from IODP-2 drilling in this area showed that slope failure has occurred multiple times since at least ~15 Ma, and that weak sediment layers appear to be associated with three separate landslide events. The observed lithological contrast of distinctly sourced sediment layers (weak diatom-derived sediments with high compressibility versus high-density glacial deposits) is proposed to have been driven by climatic events, namely changes in the extent of ice cover in the Ross Sea. Although the climate-linked layering of sediments is thought to have decreased slope stability, the trigger for landslides at this location is linked to a possible increase in the frequency of earthquakes due to rapid local uplift associated with retreating glaciers. Future warming could recreate the conditions that resulted in slope failures in the past.

**Scientific ocean drilling has also investigated the burial, storage, and cycling of carbon in ocean sediments, which has implications for understanding sources and sinks of greenhouse gases to the atmosphere.** Guaymas Basin is a young ocean spreading system in the Gulf of California that experiences very high sedimentation rates, which leads to a volcanically active rift basin blanketed by thick, organic-rich sediment layers. This unique combination results in magma-filled sills intruding into thick sediment sequences. These intrusions act as

## BOX 3.4
## Sensitivity of Earthquake Detection Using Borehole Observatories

Borehole instrument sensors, which can be installed only with scientific ocean drilling, are significantly more sensitive to subsurface fault slip movements compared with seafloor instrumentation, thus providing a higher resolution of data.

To evaluate the minimum amount of slip that can be resolved using seafloor instrumentation versus subsurface borehole instrumentation, this box includes a sensitivity test performed using a numerical model courtesy of Demian Saffer. This test compared the sensitivity of an array of eight absolute pressure gauges (APGs) to three borehole sensors. The configuration of the APGs was deployed on the seafloor at distances of 2–80 km from the trench, based on the Dense Ocean floor Network system for Earthquakes and Tsunamis (DONET) (Figure 3.14A). The three installed boreholes were within a similar distance from the trench but located much deeper under the seafloor (located 2 [C0006], 24 [C0010], and 35 [C0002] km from the trench at depths of 453, 409, and 980 m, respectively).

The numerical model (PyLITH) (Aagaard et al., 2022) incorporated realistic fault geometry, bathymetry, and variations in elastic moduli as determined by active source seismic and borehole data to simulate the response to an assumed amount of fault slip. The model simulation showed that given the noise levels associated with borehole versus seafloor sensors, the borehole sensors have the ability to resolve the fault slip <1 cm (red curve), while APGs were only capable of detecting the fault slip ≤10 cm, as shown in Figure 3.14B. When coupled with cabled observatories, borehole sensors can therefore provide significant improvement to real-time hazard assessments, including identification of slip transients leading up to major earthquakes and a more complete picture of strain accumulation and release through the earthquake cycle.

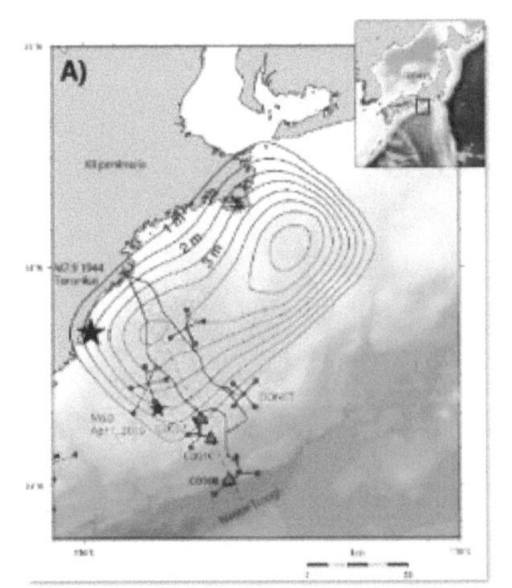


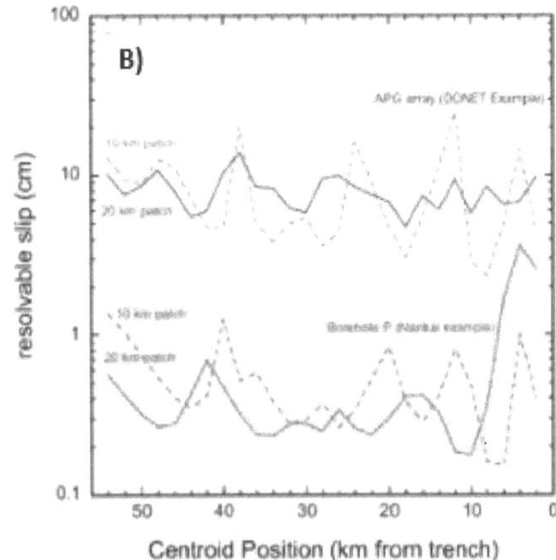


**FIGURE 3.14** (A) Contours show coseismic slip from the 1944 M7.9 earthquake (Kikuchi et al., 2014). Maroon lines indicate locations of DONET cable, and small dots indicate observatories and APGs. Red triangles indicate borehole locations. The large black star locates the epicenter of the 1944 Tonankai earthquake. The small black star shows the location of the April 2016 earthquake. (B) Models showing the amount of slip that is resolvable by the APG network (blue) and the borehole sensors (red) as a function of the centroid position of a hypothetical slip. Note that the borehole sensors can resolve approximately one order of magnitude smaller events than seafloor APGs. NOTES: APG = absolute pressure gauge; DONET = Dense Ocean floor Network system for Earthquakes and Tsunamis.

SOURCE: Figure adapted from Demian Saffer, University of Texas at Austin; map is from Araki et al., 2017. Reprinted with permission from the American Association for the Advancement of Science.

## BOX 3.5
## NanTroSEIZE: A Success Story

The Nankai Trough Seismogenic Zone Experiment (NanTroSEIZE) is a multiexpedition project by the International Ocean Discovery Program (IODP-2) to investigate fault mechanics and seismogenesis (i.e., earthquake genesis) at a subduction megathrust fault zone in the Pacific Ocean near Japan. The Nankai Trough is formed as the Philippine Sea plate subducts beneath the Eurasian plate. The megathrust fault accommodates the differential motion between the two plates and has been the site of multiple earthquakes at magnitudes of 8 or more. Systematic drilling associated with 13 IODP expeditions has resulted in direct sampling of the fault zone, as well as in situ monitoring of this megathrust fault and of overlying splay faults in the accretionary wedge of sediments (Figure 3.15A).

Through this coordinated effort, NanTroSEIZE has led to several key advances in understanding subduction zone fault systems. Prior to the NanTroSEIZE program, it was generally thought that seismogenic fault slip rarely extended up to the seafloor (Tobin et al., 2019). However, recovered fault gouge showed a highly localized fault zone that preserves evidence of thermal anomalies associated with frictional heating. These observations indicated that the fault slip extended all the way to the seafloor near the trench, significantly increasing the potential for the generation of tsunamis.

Second, borehole observatories provided new insights into the accommodation of strain on the megathrust fault. These borehole sensors provide continuous records of strain, seismicity, pore fluid pressure, and temperature. By comparing transient strain events recorded at two observatories (Araki et al., 2017), researchers were able to identify slow-slip events between 2011 and 2016, each accommodating several centimeters of slip on the plate boundary (Figure 3.15B). Collectively, these events represent 30–50 percent of the total fault slip based on the far-field plate convergence rate. Thus, the NanTroSEIZE program has elucidated that the subduction megathrust behaves in a multimode fashion, hosting both seismogenic ruptures and slow slip at different times during an earthquake cycle.

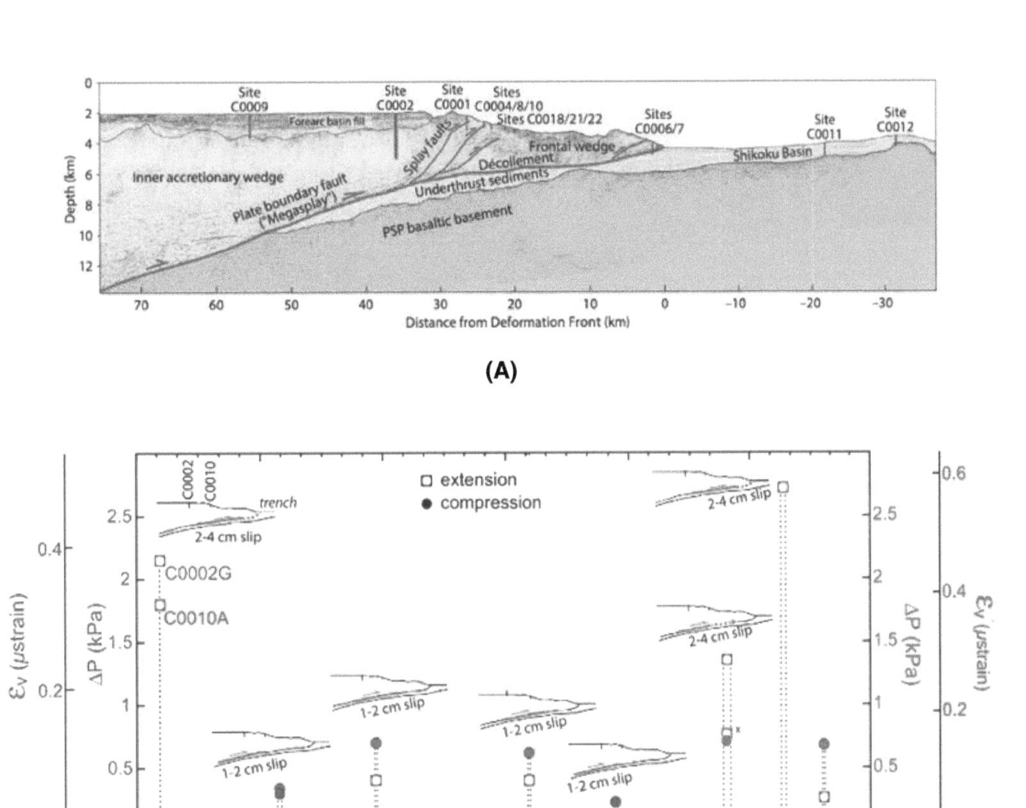


**FIGURE 3.15** (A) Cross section showing the megathrust fault (labeled as "megasplay"), marking the convergent plate boundary between the subducting Philippine Sea plate and the overriding Eurasian plate off the coast of Japan. NOTES: Smaller splay faults in this subduction zone system are also shown. Strategically placed drilling sites across this seismically active region are shown, several of which are now instrumented borehole observatories that monitor deformation in the form of strain that can be used to infer fault slip. PSP = Philippine Sea Plate. SOURCE: Tobin et al., 2015.
(B) Summary of changes in pressure ($\Delta P$) and strain ($\varepsilon_v$) measured at two boreholes (red and blue) near the Nankai trench off the coast of Japan. NOTES: Some motion was compressional (solid circles), while other motion was extensional (open squares). Dashed vertical lines indicate duration of each event. kPa = kilopascal. SOURCE: Araki et al., 2017. Reprinted with permission from the American Association for the Advancement of Science.

transient heat sources that drive off-axis hydrothermal circulation with the potential to release sedimentary carbon as methane. IODP-2 drilling in the Guaymas Basin investigated these processes and found that the sills not only provide the heat necessary to release hydrocarbons, but also act as chemical reaction zones sequestering some carbon into the sill matrix (IODP, 2019). Furthermore, the structure of young rift basins (i.e., individual rift segments with different elevation depocenters) controls the distribution of sediment deposition relative to variations in sea level. Additionally, drilling results from scientific ocean drilling in the Gulf of Corinth suggest greater carbon burial during interglacial periods, when deposition occurred under marine conditions, compared with glacial periods, when the basin was closed off from ocean. The mechanism relates to local controls of sea level on the carbon cycle: during warm interglacial periods, this location was below sea level and a better place for depositing and burying carbon. But during glacial periods, the sea level was lower, and it was a less productive, closed-off setting, which resulted in less carbon burial.

*High-Priority Future Research*

Scientific ocean drilling can enable direct sampling of fault rocks, which allow scientists to infer the material properties of the active fault zone, most importantly the "megathrust" (i.e., the fault that separates the downgoing tectonic plate from the overriding tectonic plate). **A key outstanding question regarding the behavior of megathrust faults is under what conditions they remain "locked" (i.e., accumulating stress without slipping up until a major earthquake), and/or slip via aseismic creep, releasing stress slowly and preventing the stress accumulation that leads to large earthquakes.** The difference between these two types of fault behavior translates directly into the seismic hazard that will be experienced at a specific subduction zone and can be tested through scientific ocean drilling. Proposed mechanisms to explain these different behaviors include variations in fault thermal structure, rock composition, and/or pore fluid pressure. To probe the behavior of the megathrust, samples of fault material recovered in cores can be brought back to the laboratory, where their frictional properties are tested experimentally. Performing experiments on natural fault rocks, as opposed to synthetic samples, is essential, because it assures that the experimental material is of the same composition as the actual fault zone.

Future studies at other subduction systems (e.g., Cascadia, Alaska, Chile, and the Caribbean) will allow scientists to better understand different conditions (e.g., fault zone composition, fluid pressure) that promote either seismogenic or stable (non-earthquake-producing) fault motion (i.e., aseismic creep). If such records were obtained, these data could be used to constrain numerical models of dynamic fault ruptures, earthquake cycles, and tsunami genesis, such as those models being developed by the NSF-funded Cascadia Region Earthquake Science Center.

**CONCLUSION 3.3c** Future studies of subduction systems, including borehole monitoring, will allow scientists to better understand different conditions that promote either seismogenic or stable (non-earthquake-producing) fault motion. These data will constrain numerical models of dynamic fault ruptures, earthquake cycles, and tsunami genesis to advance understanding of the conditions under which natural hazards occur and to create a more robust warning system.

## Exploring the Subseafloor Biosphere

*Advancing understanding, discovery, and characterization of the world of living microbes below the seafloor.*

Comprising bacteria, archaea, viruses, fungi, small eukaryotes, and even small invertebrates, the subseafloor biosphere concerns life existing in sedimentary, crustal, and fluid environments deeper than 1 m below the seafloor sediment–water interface (Orcutt et al., 2013, and references therein), potentially extending for hundreds of meters beneath the seabed. Evidence of living biomass has been observed as deep as 2.5 km (Jørgensen and Marshall, 2016) below the seafloor, with about 80 percent of all archaea and bacteria on Earth predicted to reside in the subsurface (Figure 3.16; Bar-On et al., 2018; Kallmeyer et al., 2012). Figure 3.16 indicates that the deep subsurface may contain approximately 12 percent of the total global biomass. A variety of habitats exist in the

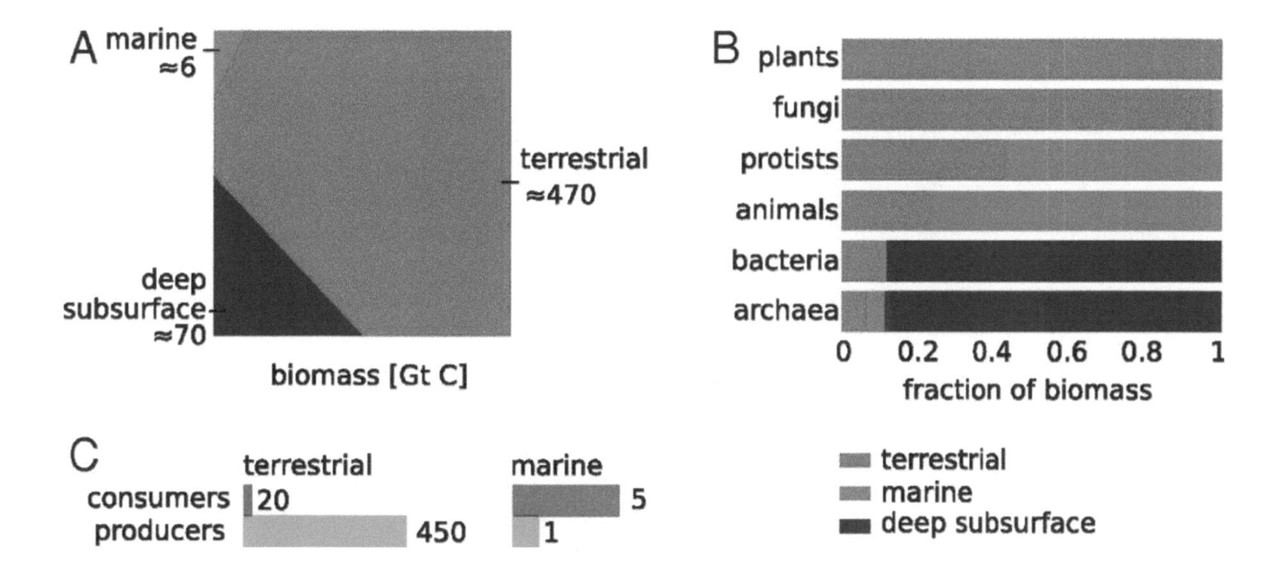


**FIGURE 3.16** Within the vast expanse of the oceanic subseafloor biosphere, gigatons of deposited and buried carbon (Gt C) and an estimated $10^{27}$–$10^{29}$ cells are present. Biomass is distributed across different environments and trophic modes. (A) Absolute biomass shown using a Voronoi diagram, with the area of each cell being proportional to the global biomass at each environment. (B) Fraction of the biomass of each kingdom concentrated in the terrestrial, marine, and deep-subseafloor environments. (C) Distribution of biomass between producers (autotrophs, mostly photosynthetic) and consumers (heterotrophs without deep subsurface) in the terrestrial and marine environments. The size of the bars corresponds to the quantity of biomass of each trophic mode. Numbers are in gigatons of carbon.
SOURCE: Bar-On et al., 2018. https://creativecommons.org/licenses/by-nc-nd/4.0/.

subseafloor biosphere: sediment, pore water (interstitial water), crust, fluids that circulate within the crust, and in isolated seafloor habitats, such as hydrothermal vents and methane seeps.

The study of the subseafloor biosphere includes the relationships and interactions within microbial communities and their surrounding environment. Sediment cores collected kilometers deep reveal a remarkable diversity of microorganisms and surprising metabolic complexity, as well as an overarching biological imprint on the chemistry of the overlying sediments and the ocean.

The deep-ocean subsurface biosphere is far removed, spatially and temporally, from photosynthetic productivity. Accordingly, this energy- and nutrient-limited habitat exhibits ecological, physiological, and evolutionary patterns and processes that are markedly different from those of other habitats (Hoehler and Jørgensen, 2013). Fed by only the organic content of deep-sea sediments, living cells in the subseafloor biosphere have low metabolism and long division times. As they are buried by organic sediment accumulation, nutrient availability decreases and metabolic rates (and thus growth) slow (Jørgensen and Marshall, 2016). In addition, sediments where temperatures rise above 50°C also support microbial communities, where the physiological and biochemical adaptations to low energy at elevated temperatures remain unknown, representing a vibrant area of research. On the whole, the conditions found in the subseafloor biosphere—which include limited bioavailable nutrients and substrate for energy conservation, extremes in temperature and pressure, variable redox conditions, and even biophysical challenges such as minimal porewater volumes that constrain cell size—have resulted in a myriad of adaptations that confer greater fitness in these environments at the edge of the biosphere.

Although the deep-subsurface biosphere contends with limited nutrient availability and chemical and physical challenges, it has in recent years become apparent that the subsurface microbial community may be exerting a marked influence on ocean biogeochemistry. For example, it has been shown that the ocean contains one of the largest pools of reduced organic carbon on Earth (Druffel et al., 1992; Hedges, 1992). Some of this carbon seems to persist for millennia and is considered recalcitrant (i.e., resistant to biological or abiotic breakdown). Nevertheless, there is a measurable loss of organic carbon, including recalcitrant organic carbon, that cannot be accounted for by abiotic removal, such as particle adsorption and burial. Over the last decade, IODP and NSF-supported research at the North Pond deep-subsurface study site in the mid-Atlantic have shown that ocean bottom water can circulate quickly through the subsurface aquifer over relatively short timescales. Moreover, through a series of isotopic geochemical measurements, molecular biological studies, and live microbial incubations, studies have shown that microbial communities in the subsurface biosphere are playing a quantitatively relevant role of recalcitrant organic carbon removal (Reese et al., 2018; Shah Walter et al., 2018; Trembath-Reichert et al., 2021). Some members of the deep subsurface biosphere are also chemosynthetic, generating new biomass and labile organic carbon in the deep subsurface that may be playing a key role in sustaining this ecosystem. These and other studies were groundbreaking in that they quantified the metabolic rates of deep-subsurface microbes and revealed their net impact on organic carbon degradation and production. In sum, these studies underscored the critical but poorly constrained role that subsurface seawater circulation plays in the ocean carbon cycle.

The conditions of the subseafloor biosphere have placed tremendous evolutionary pressure on the microbes that reside therein, yet it is clear that the extant biosphere is active (to some degree), diverse, and so abundant that they possibly exceed the total number of cells in the overlying water column (Jørgensen and Marshall, 2016). The results of previous and current studies illustrate that this community harbors great potential for the discovery of novel microorganisms and/or new metabolic capabilities. They also are likely to play a significant role in global biogeochemical cycles (Bar-On et al., 2018; Lever et al., 2013). Yet to date, understanding of their diversity, evolution, adaptations, and resilience (e.g., response to perturbations) is very limited. **Understanding the diversity, abundance, and metabolic scope of subseafloor microbes, their role in marine biogeochemical cycles, and their sensitivity to natural and anthropogenic disturbances (including increasing mean ocean temperatures and deep-sea mining) are critical to current understanding of this habitat that harbors the majority of marine microbes.** Microbiological sample collection during scientific ocean drilling efforts has occurred since the 1960s, but to a limited extent, and in its earliest days was relegated to cell counts of living and dead organisms. Beginning in the 1990s, concurrent with the development of next-generation microscopy and "-omics" techniques, the ability to collect and preserve samples for microbiological analysis was prioritized during drilling expeditions. To enable these studies, the community developed sampling and preservation techniques to prevent and monitor contamination, enable preservation for DNA- and RNA-based analytical methods, and provide enough relevant samples to measure microbiological activity and for cultivation experiments. This work established sampling protocols that were incorporated into scientific ocean drilling, enabling the description of the diversity, abundance, and metabolic activity (rates) of subsurface microbiomes from cores.

Over the last two decades, massive and rapid advances in genomic sequencing and bioinformatics have allowed researchers to explore microbial genomes and functions and to conduct comparative genomic analyses. This analysis has provided many new insights into the functionalities of the diverse microbes found in the subsurface and has revealed that energy, nutrient availability, and the physical environment are important drivers of microbial communities in subsurface environments, yielding additional insight into metabolic flexibility, adaptability, and energy dynamics of these organisms in their extreme environments.

Additionally, seafloor and subseafloor experimental apparatus designed to facilitate research on living, metabolically active microbes have also advanced subseafloor biosphere research. This includes installing observatories for deep-crustal environments using scientific ocean drilling and incorporating continuous fluid-sampling installations in drilled borehole casings. These installations provide more direct means to measure microbial activity in the deep biosphere, coupled with in situ biogeochemical analyses to further elucidate the importance of microbial activity and presence in the subseafloor biosphere.

### *Progress Made During IODP-2*

**Scientific ocean drilling has conclusively demonstrated two key outcomes on life within the subseafloor biosphere: (1) its abundance and diversity and (2) the activity within this subseafloor biosphere.** Cores of ocean subseafloor sediments have found a large diversity of microbial life living kilometers below the sediment surface, including in sediments dated over 100 myr old (Jørgensen et al., 2020). There is a tremendous diversity of microbial life in these sediments, including a large range of phyla in archaea and eubacteria. Cultivating microbial life is very difficult for these microbes, which have generation times of years to thousands of years in the subsurface, and much of what is known about their lives comes from sequencing-based surveys of gene function and taxonomy. In addition to microbes, researchers have found surprising abundance and diversity of fungal and viral groups, even metazoans, such as bacteria-feeding nematodes.

Understanding of the subseafloor biosphere was greatly expanded during IODP-2, with expeditions dedicated to collecting and studying living microbes in ocean sediment and rock. The chemistry of pore water collected from sediment cores provided the first evidence of the possibility of life below the sediment surface. Furthermore, direct counts of bacteria provided confirmation of life deep within these subsurface sediments (e.g., Parkes et al., 1994). Expeditions in subsequent years revealed that oceanic crust also contained the presence of metabolically active microbes (D'Hondt et al., 2019a). A series of studies from the mid-Atlantic also revealed the presence of relatively rapid circulation of deep oxygenated bottom water through the basaltic aquifers of the deep subsurface (Edwards et al., 2014), and evidence of both recalcitrant carbon degradation and carbon fixation by the endemic microbes (Shah Walter et al., 2018; Trembath-Reichert et al., 2021), indicating that deep-subsurface communities play a role in deep-ocean carbon cycling. More recent studies of sediments and crust, including midocean ridge flanks and undersea volcanoes, revealed the presence and activity of diverse microbial communities living deep within sediments and the crust and in low- and high-temperature settings (up to 120°C; e.g., Beulig et al., 2022; Heuer et al., 2020; Li et al., 2020; Reysenbach et al., 2020). Within this subsurface biosphere, where sediments older than 100 myr exist at 2.2 km below the seafloor, a possibility of life has been demonstrated, surviving at energy levels far below those previously thought to be the limit, and with metabolic rates many orders of magnitude slower than those seen in biospheres on Earth's surface (D'Hondt et al., 2019a). **This research has opened new horizons of knowledge into the basic building blocks of life, which may correlate to understanding of the potential for life in other areas of the solar system (and beyond), the origins of life on Earth, and the integral building blocks of ecosystems that nurture the biological world.**

In addition to leading to the potential understanding of unknown worlds, the subsurface biosphere has significant impacts on ocean biogeochemistry, since its metabolic products diffuse or circulate toward the surface and eventually make their way into the overlying ocean waters. As mentioned, studies have implicated deep-subsurface microbes in deep-ocean carbon cycling, and it has also been shown that they influence the nitrogen, sulfur, and other elemental cycles. For example, the metabolism of subseafloor microbes affects many chemical fluxes to the seafloor, including the burial rate of organic matter and production of pyrite, which is the principal sink for sulfur in the ocean (Figure 3.17). The production of pyrite is a principal source of alkalinity for the ocean, affecting the exchange of $CO_2$ between the ocean and atmosphere. The metabolism of the subseafloor biosphere also reduces the amount of nitrate nitrogen available for ocean primary production and plays an important role in creating and/or destroying economically significant deposits of hydrocarbons, phosphate, dolomite, and barite (D'Hondt et al., 2019b). Figure 3.17 shows subseafloor microbial activities influencing chemical fluxes in the ocean seafloor and ocean.

Efforts to study the subseafloor biosphere are benefiting from advances in the ability to characterize microbial life forms using DNA- and RNA-based analytical approaches, advances in defining the specific geochemical reactions that enable microbial metabolism, and advances in experimentation that enable tracking microbial metabolism. Indeed, results from recent scientific ocean drilling demonstrate that microbes are present under a wide range of energy conditions, but the seafloor is characterized by modest to high hydrothermal fluxes and water–rock reactions, which tend to promote the most diverse microbial communities (Reysenbach et al., 2020). These are also settings where inorganic precursors to life may have provided an environment in which early life forms could have originated and/or evolved. Based on metabolic calculations, some recent studies have concluded that cells in the subseafloor biosphere may be able to survive for hundreds to thousands or perhaps even millions of years, not by dividing to produce new cells, but by using the small amount of available energy to repair the key components of cellular life (D'Hondt et al., 2014). The literature also indicates mortality as an important aspect to life in the deep biosphere (Jørgensen and Marshall, 2016), with evidence of turnover of dead microbial biomass (Lomstein et al., 2012), potentially facilitated through viral lysis (Cai et al., 2019).

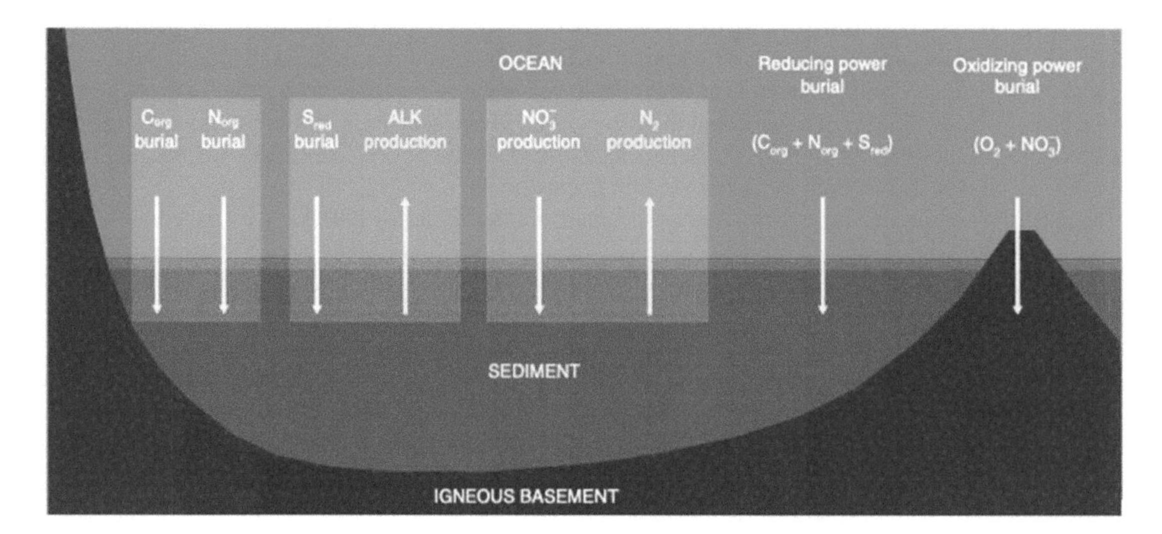


**FIGURE 3.17** Major net chemical fluxes due to microbial activities in subseafloor environments. Subseafloor microbial activities control burial fluxes of organic carbon ($C_{org}$), organic nitrogen ($N_{org}$), and reduced sulfur ($S_{red}$) (typically buried as iron sulfide), as well as production fluxes of alkalinity (ALK) and $N_2$. In doing so, they control the rate at which reducing power is buried. Microbes in igneous basement contribute to burial of oxidizing power (via reduction oxygen and nitrate).
SOURCE: Figure and caption from D'Hondt et al., 2019b.

*High-Priority Future Research*

**Key topics that remain in subseafloor biosphere research focus on understanding the limits to life; the way biological communities interact, move, and evolve within the subsurface biosphere; and their distribution across space and time in the subsurface environment**. Measurements of microbe metabolism are primarily from sediments; less is known of the activity in the ocean crust. Three fundamental and compelling questions have yet to be answered: (1) How do cells survive for very long periods of time with minimal nutrition and substrates for energy conservation, and what is the role of mortality in survival? (2) How do microbes move around or otherwise pass genetic information throughout the deep-subsurface biosphere, and what are the implications for their evolution? (3) To what extent does the deep-subsurface biosphere play a role in marine biogeochemical cycles, including carbon capture, carbon sequestration, nitrogen fixation, bioavailable nitrogen reduction, etc.? Furthermore, understanding how microbes regulate and control energy flow in extremely energy-limited subseafloor environments will provide insight into the potential for life on other planets and into the evolution of organisms in relation to their environments. Additionally, **the role of subseafloor microbial life in the global carbon budget, and its associated feedbacks, will remain largely unexplored without additional scientific ocean drilling**.

Addressing these priorities has been aided by advances in ocean drilling protocols that enable relevant microbiological research, such as the ability for cryopreservation of core samples and improved access to samples for cultivation and activity experiments. The subseafloor biosphere community has consistently called for standardization of methods across drilling platforms to better enhance the research efforts. Basic microbial measurements are now included on some drilling legs, primarily to enumerate and identify which organisms are present. **Measurements of microbe metabolic activity continues to be elusive based on limited sample materials and the inability to make measurements of biological activity on frozen or preserved samples**.

**CONCLUSION 3.3d** Scientific ocean drilling is necessary to address key unanswered questions about the subseafloor biosphere and to advance understanding of the limits to life, as well as the way biological communities interact and move within the subsurface biosphere and how they are distributed across space and time. Such research has direct implications for understanding the potential for life in other areas of the solar system, the origins of life on Earth, and the integral building blocks of ecosystems that nurture the biological world.

## Characterizing the Tectonic Evolution of the Ocean Basins

*Advancing understanding of the dynamics of tectonic processes and the cycling of energy and matter between Earth's interior and surface environments.*

Earth is a dynamic system because of its plate tectonics. The overturning of the mantle and formation and destruction of tectonic plates is a constant source of energy and matter, thus impacting virtually every Earth system. Ocean basins encompass the life cycle of oceanic plates, from the generation of new oceanic crust at mid-ocean ridges to the ultimate subduction of plates at convergent boundaries (Figure 3.18). The dynamic functions of this planet cannot be understood without understanding the driving forces that control the birth and death of oceanic plates. Furthermore, the ocean basins and midocean ridges are a vital source of geologic resources (e.g., rare metals and other essential minerals). There is also increasing interest in utilizing the ocean crust as a carbon storage solution.

Every ocean basin started as a continental rift, with upwelling mantle material driving continental extension and breakup, or in some cases, as a preexisting weakness in the crust driven by extensional forces. This process can be accompanied by the eruption of flood basalts representing a large and dramatic input of matter (including greenhouses gases) and energy, which have been linked to mass extinction events. **These initial, dramatic stages of ocean creation often can be evidenced only by drilling through up to a kilometer or more of sedimentation.** Scientific ocean drilling can provide vital information on the structure and thermal evolution of ocean margins that are formed during the rifting process and serve as a significant repository of energy resources.

Once continental rifting evolves into oceanic spreading, midocean ridges become the engines of oceanic crustal formation and a critical site for transfer of energy and matter from the mantle to crust and overlying ocean. Midocean ridges create two-thirds of Earth's surface and drive its repaving approximately every 100–200 myr. Midocean ridges also serve as the site of unique, local chemosynthetic ecosystems, and, through their dynamic geochemical exchanges

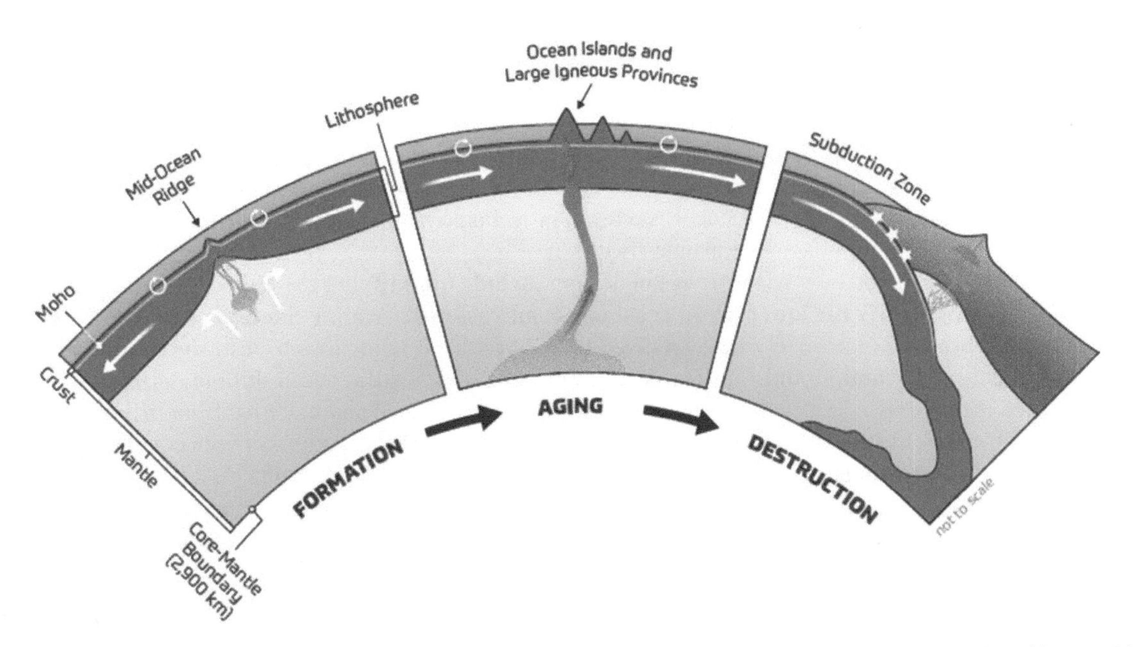


**FIGURE 3.18** Three main phases in the life cycle of an ocean plate. NOTES: Phases include formation at the midocean ridge (left), aging as it moves away from the midocean ridge (middle), and destruction at a subduction zone (right). The oceanic plate is dark gray, continental lithosphere is brown, and a mantle plume (a hotspot) is rising up from the core–mantle boundary. White arrows show plate motion, white circular arrows depict hydrothermal circulation, and white stars are large earthquakes. SOURCE: Koppers and Coggon, 2020. Illustration by Geo Prose.

with the overlying ocean, help enable life on Earth globally. Midocean ridges range in spreading rate from ultrafast (~15 cm/yr) to ultraslow (fraction of a centimeter per year), producing distinctly different types of seafloors. In addition, spreading rates vary through time, and past spreading rates are thought to have been even faster than current rates. The fastest spreading rates are driven by subduction at the margins of ocean basins and produce a topographically high and relatively smooth seafloor. In slower spreading rates, the lithosphere is cooler, and topography becomes more jagged, leading to a deep axial valley bounded by large normal faults. Ocean drilling has helped ground truth these radical differences in volume, composition, and architecture of the crustal structure.

Hydrothermal circulation is most pronounced at and near the ridge crest, fueled by the central magma supply and active faulting, where black smoker vents emit geochemically laden fluids, at temperatures sometimes in excess of 400°C. This thermochemical exchange is both a sink and source of energy and matter with the overlying ocean, although the exact balance of this is poorly understood, as fluids enter and exit the crust in both focused and diffuse sites. This exchange has significant impacts on the overlying ocean and atmosphere, serving as both a sink and source of greenhouse gases and ocean chemical variability. As the crust ages, hydrothermal circulation can remain an important source of cooling in crust as old as 65 myr. The threshold between purely conductive and advective cooling is important to understand, for not only fundamental plate processes, but also the extent and nature of the geochemical exchange. Serpentinization is one manifestation of this hydrothermal exchange and is particularly pronounced at slow and ultraslow ridges where the scale of faulting is larger. Serpentinization has other intriguing roles—it affects abiotic organic carbon availability in hydrothermal systems and opens new pathways for organic synthesis (Andreani et al., 2023). **Such processes are predicted on other planets and/or moons, and thus inform comparative planetary geology and astrobiology**.

The ocean is constantly undergoing events that take place after the initial formation of new crust at the ridge crest. From small seamounts to hotspot islands to flood basalts, volcanism punctuates every ocean basin, always revealing insight into the complex workings of the mantle beneath. Given that ocean crust covers 70 percent of Earth's surface and lies beneath all oceanographic phenomena, scientific ocean drilling has an important role to play in understanding these geologic processes.

*Progress Made During IODP-2*

Since 2013, scientific ocean drilling has made progress in addressing questions related to better understanding the structure, genesis (spreading), and destruction (subduction) of ocean crust, the nature of the boundary with the upper mantle, and the overall influence of fluid flow through the crust on the chemical composition of the ocean and atmosphere. Some of the progress made was a result of technological advances that enabled drilling through the crust and up to 1.5 km into the upper mantle (Box 3.6).

**Key contributions** to date by scientific ocean drilling **include quantifying the tectono–magmatic interactions that form and modify the lower ocean crust on the ultraslow-spreading Southwest Indian Ridge, and finding that gabbros** (i.e., intrusive igneous rock), which crystallized in the lower crust, **were being modified by crystal–plastic deformation and faulting** (Dick et al., 2019). Scientific ocean drilling during IODP-2 has also elucidated the processes by which ocean crustal architecture is created and modified from rifting to seafloor spreading. For example, drilling in the South China Sea determined that the transition between continental breakup and igneous seafloor spreading occurred quickly—within 10 myr (Larsen et al., 2018). Additionally, cores from the western Pacific Izu–Bonin–Mariana subduction zone yielded evidence to support modeling the spontaneous initiation of plate subduction (the subsidence of dense lithosphere along faults adjacent to buoyant lithosphere [Arculus et al., 2015]).

**Scientific ocean drilling has also developed new insights and informed models for chemical and fluid exchanges between seawater and ocean crust**. At any one time, approximately 2 percent of seawater is moving though volcanic rock exposed at midocean ridges or residing below sediments, implying that the entire volume of the ocean cycles through the subseafloor basement about every 200,000 years. While flowing through the basement rock, the seawater is altered by microorganisms and water–rock reactions, affecting ocean chemistry. For example, results from drilling ultradeep sites in the Japan Trench indicated microbial-mediated dynamic carbon cycling (Chu et al., 2023). Results from drilling in the Mariana and Izu–Bonin subduction zone system demonstrated

**BOX 3.6**
**Fulfilling a 60-Year Goal of Scientific Ocean Drilling**

For nearly 60 years, teams of geoscientists and ocean engineers have worked together on a scientific and technological challenge: to drill and recover core through the Earth's crust and into the underlying mantle rock. Obtaining an intact, long, and continuous sequence of ocean crust and mantle rock would provide tangible evidence of the structure and composition of Earth's interior. Such material would be extremely special, as most studies of Earth's interior must depend on indirect means, such as seismic data from earthquakes or altered igneous rock samples that have been caught up in the convergence of tectonic plates and emplaced on the continents. Achieving this objective would push the limits of technology and innovation and require decades of tool development and testing to overcome the challenging temperature and pressure conditions of Earth's interior, to which the drilling tools would be subjected during the drilling and core-recovery process.

In 2023, on IODP Expedition 399, this long-awaited goal was accomplished, marking a major benchmark in Earth and ocean sciences. Expedition 399 took advantage of a location on the seafloor along the Atlantic midocean ridge, where the ocean crust was thin, creating a "tectonic window" into the underlying mantle. Here, the expedition drilled an unparalleled record 1.5 km into Earth's upper mantle, recovering cores of grey-green rock (Figure 3.19). Following core recovery, additional data were collected by running geophysical logging tools into the drilled hole. Core samples and logging data are now being used together to better understand the nature and dynamics of the tectonic plate at this location, and how fluids and elements are exchanged between the lithosphere and ocean waters. Samples of organic molecules and microbes were also collected from the crustal and mantle rocks to study the controls on life deep beneath the seafloor, and to test hypotheses on the origins of life on Earth. This scientific and engineering accomplishment demonstrated that recovery of such records is possible, providing the knowledge necessary for future drilling operations in a variety of tectonic settings below the ocean.

**FIGURE 3.19** Cores of mantle rock from below the seafloor recovered from Expedition 399. Toothpicks with green stickers mark places where scientists plan to sample for post-expedition research.
SOURCE: Johan Lissenberg, Cardiff University, IODP.

that anomalies in the marine silica budget may be explained by low-temperature chemical reactions with seafloor basement rocks (Geilert et al., 2020). Additional research has contributed to understanding the roles of fluids in triggering volcanic eruptions. Based on drilling in the Mariana arc, a new model for processes of dehydration of subducting plates and magma production emerged based on elemental proxies for chemical exchanges (H. Li et al., 2021).

*High-Priority Future Research*

To date, only ~45 km of crustal cores have been collected via scientific ocean drilling, and less than 100 holes have been drilled that extend deeper than 100 m into basement (crustal igneous) rock. Furthermore, understanding of ridge processes is dominated by materials from only seven cores collected on active ridges. **While IODP-2 has made stepwise progress to develop models for fast- to slow-spreading ridges, the models can be tested only by obtaining cores from other ridge locations.**

Additionally, a vital method for ground truthing the geochemical impact of hydrothermal circulation is by drilling and recovering hard-rock cores, which can provide in situ samples of the alteration taking place in a variety of settings. Equally poorly understood is how the process of serpentinization impacts biogeochemical cycles. Developing large-scale spatiotemporal records of serpentinization is important for understanding the carbon cycle, as well as mass balance discrepancies and gaps in other global elemental budgets. **Better understanding of the processes (e.g., exchanges of fluids with subseafloor materials), geographies (e.g., hydrothermal vent settings, underwater volcanoes, trench sediments), and other variables that influence elemental mass balances would be fundamental, important new contributions of scientific ocean drilling and borehole observatories that have implications for global oceanography**. Understanding serpentinization more broadly is also important for understanding potential geohazards, as it can weaken and lubricate faults.

The midocean ridge environment provides a relatively shallow location in which to study these questions, and scientific ocean drilling has a vital role to play. Just as natural hydrothermal circulation deposits minerals deep in the oceanic crust, spurring interest in deep-sea mining, there is also increasing interest in the ocean as a potential reservoir for carbon dioxide (Box 3.7). Understanding both the natural inorganic calcium carbonate precipitation and consequences of storing additional carbon are topics that can be better quantified by examining ocean crust recovered from a variety of locations.

During Deep Sea Drilling Project 2, drilling to understand subduction zone processes was dominated by sites near Japan and South Asia. **New research is needed to improve the global understanding of important physical and chemical processes at subduction zones.** This includes more information on how mantle melting processes (that feed volcanoes on overriding plates) evolve during and after subduction initiates, how geometric complexities and heterogeneities within fault zones impact subduction; and how methane sources and sinks in subduction zone hadal trenches impact the global carbon cycle.

**CONCLUSION 3.3e** Only scientific ocean drilling can provide key constraints regarding the formation and evolution of oceanic crust and the upper mantle. The cycling of fluids through the subseafloor and corresponding chemical exchanges between the liquid and solid Earth have implications for processes with direct societal relevance, including the production of mineral resources, sequestration of atmospheric carbon dioxide, and origin of geohazards (including volcanic eruptions, earthquakes, and related tsunamis).

## BOX 3.7
## Scientific Ocean Drilling and Ocean $CO_2$ Removal and Sequestration

The ocean is Earth's largest carbon sink, presently absorbing an estimated 25 percent of all $CO_2$ emissions. By accelerating and supplementing natural oceanic processes, including those related to subsurface burial of carbon, the ocean may play an even bigger role in absorption and long-term carbon storage, as a component of a larger climate mitigation strategy. Several strategies for increasing the ocean's capacity to absorb and sequester $CO_2$ are being considered, and small-scale experiments are underway. However, fundamental questions remain, which impede the scale-up of any potential solution, including unknown and unintended environmental consequences and efficacy of any particular approach (NASEM, 2022b). Scientific ocean data may be able to provide insight into these questions. In particular, it can play an important role in characterizing the chemical and physical variability of basement rock (including stress and pressure conditions), in understanding rock alteration and mineralization, and in evaluating the potential for carbon sequestration in the ocean crust. There are some indications that offshore basalts may be a durable location for carbon sequestration because injected $CO_2$ is expected to mineralize, forming carbonate rock (Ekpo Johnson et al., 2023; Goldberg and Slagle, 2009). Existing scientific ocean drilling locations and data can help evaluate such conditions. For example, logging data from a suite of Ocean Drilling Project and International Ocean Discovery Project sites in the Cascadia Basin have been used to analyze fault slip that might affect a potential deep-sea balsalt $CO_2$ injection site (Ekpo Johnson et al., 2023). Additionally, Expedition 392, coring a submarine plateau near South Africa, retrieved a subseafloor record of enhanced basalt weathering that could provide a natural laboratory for investigating the impacts of proposed climate mitigation techniques. Expedition 396 recovered layered sequences of vesicular volcanic and sedimentary rock off the coast of Norway that offers another opportunity to assess subseafloor characteristics needed for $CO_2$ sequestration. Existing deep-ocean drilling data, new hard-rock recovery and analyses, and borehole observatories (to monitor, e.g., future sequestration experiments; Figure 3.20) will be important for evaluating the subseafloor environments for potential carbon storage.

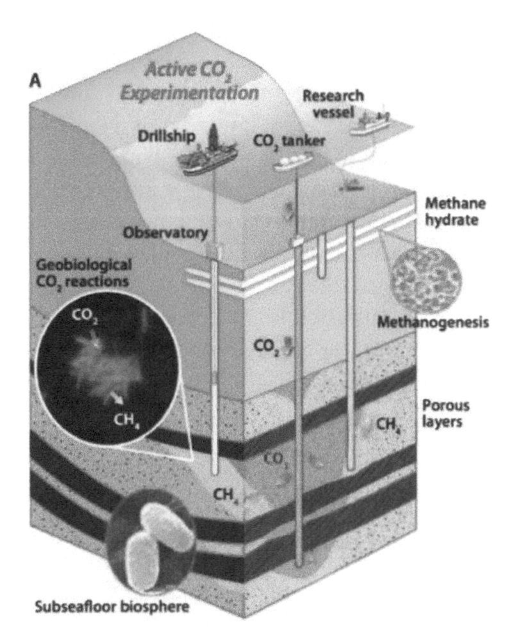


**FIGURE 3.20** Carbon dioxide sequestration experimentation.
SOURCE: Fumio Inagaki, JAMSTEC.

## CLASSIFICATION OF SCIENTIFIC OCEAN DRILLING–SUPPORTED RESEARCH

Important scientific research advances a field of study in some way. However, not all important scientific research is also vital and/or urgent. The following definitions apply to the committee's evaluation of future high-priority research requiring scientific ocean drilling for advancement:

> ***Vital science*** encompasses compelling, high-priority research with the potential to transform scientific knowledge of the interconnected Earth system and the critical role of the ocean in that system. Vital scientific research can lead to paradigm shifts in understanding, potentially opening new doors to research and technology innovations that can benefit humanity with direct societal relevance.
>
> ***Urgent science*** is time sensitive and has immediate societal relevance to emerging challenges at regional to global scales. It is scientific research that needs to be done now in order to understand changes or new circumstances that can inform predictive models and decision making and may be related to tipping point vulnerabilities. It implies that immediate action is needed and thus is a higher priority than vital science.

Elevating a research area to one or both tiers is not done lightly. Recognizing that the terms above label complex issues, the committee has carefully described their working definitions. The committee has also tried to ensure that the labels are not undermined by overuse. For example, almost all basic research can potentially lead to paradigm shifts in understanding in the future; this has long been the reason for NSF's commitment to basic research. But in the committee's use of the term, *vital* is reserved for work perceived to be at the cusp of major shifts in scientific understanding. It recognizes the serious nature of regional and global change, as well as the areas of greatest need and potential impact requiring scientific ocean drilling. The committee recognizes that funding for scientific research is not unlimited and forward-looking prioritization is needed to guide investments in research, infrastructure, and workforce development. It also recognizes the essential role of scientific research, and supporting related facilities and technology, in U.S. leadership on the global stage.

Each of the five high-priority scientific areas that require future scientific ocean drilling for advancement are determined to be important and vital science. There was strong committee consensus that two of these research areas are also urgent science. For one area, the committee had a broader range of views, and for this reason it is categorized as *potentially urgent* (indicated with parentheses around the check mark in Table 3.2).

**Each of these five high-priority research areas is vital to advancing scientific understanding of the interconnected Earth system and has the potential to lead to paradigm shifts in its field.** In particular, the exploration of subsurface microbial life is on the cusp of major discoveries that are expected to transform the field. Subseafloor sediment and hard-rock records are essential to understanding what makes the planet habitable, and where and how life originated and evolved. It requires knowledge of the complex exchanges of fluids and nutrients that occur between the subseafloor biosphere, Earth's crust, the ocean, and the atmosphere. Sampling Earth's oceanic crust at different ages all around the planet provides insight into the processes that govern the occurrence of earthquakes, tsunamis, and volcanoes and the global cycling of energy and matter that produces critical economic resources. Additionally, long, continuous, and high-resolution paleoclimatic and paleoceanographic sedimentary records from

**TABLE 3.2** High-Priority Science Areas that Require Future Scientific Ocean Drilling

| Future Scientific Ocean Drilling Priority Science Areas | Vital Science | Urgent Science |
|---|---|---|
| Ground Truthing Climate Change | √ | √ |
| Evaluating Marine Ecosystem Responses to Climate and Ocean Change | √ | (√) |
| Monitoring and Assessing Geohazards | √ | √ |
| Exploring the Subseafloor Biosphere | √ | |
| Characterizing the Tectonic Evolution of Ocean Basins | √ | |

the subseafloor are vital to constrain the processes that regulate or destabilize feedbacks in Earth's climate system and to examine the geological record of past tipping points and transient climate states, and the dynamics of ice–ocean–atmosphere interactions in past periods of elevated temperatures.

Research on changes in microfossil assemblages in the distant past in response to environmental changes—such as ocean circulation, pH, temperature, and oxygen content—are prophetic in that they may indicate possible future changes in ocean ecosystems. Many studies and models of modern ocean ecosystems seek to understand how changes in ocean properties could affect marine species and ecosystems in the 21st century. Research about the past, such as that based on sedimentary microfossils, is vital, and also becomes urgent to understanding the potential response of modern ocean ecosystems to rapid changes in conditions during the next 100 years. Changes in the ocean food web could trigger declines in species upon which many coastal communities depend for sustenance and economic stability.

*Ground truthing climate change* and some aspects of *evaluating marine ecosystem responses to climate and ocean change* are deemed urgent because it is only from the records recovered through scientific ocean drilling that analogs for modern and near-future challenges of rapid global warming, sea level rise, and widespread ocean acidification and deoxygenation can be observed. Data from these records are needed to inform predictive models today. Borehole observatories, which can only be installed through scientific ocean drilling, are used for what is considered urgent research. They are needed for identifying precursory earthquake events and are an order of magnitude more sensitive to fault slip than other real-time systems (e.g., seabed observations), providing better data records.

**CONCLUSION 3.4** *Ground truthing climate change* and *monitoring and assessing geohazards* are both identified as urgent and vital research priorities critical to advancing U.S. national interests. *Evaluating marine ecosystem responses to climate and ocean change* is identified as vital and potentially urgent. *Exploration of the subseafloor biosphere* and *characterizing the tectonic evolution of the ocean basins* are identified as vital research.

## CONNECTIONS TO NATIONAL PRIORITIES AND PREVIOUS RECOMMENDATIONS TO NSF

The five high-priority areas that require future scientific ocean drilling for advancement are connected to national priorities, as well as recommendations made to NSF in previous reports published by the National Academies of Sciences, Engineering, and Medicine (e.g., NRC, 2015; NASEM, 2022a,b,c) (Table 3.3).

Planning for and forecasting a future Earth requires not only looking at historical records but also using the data to feed into future predictions. A good understanding of ocean ecosystem responses and climate history, provided by scientific ocean drilling and supplemented with other oceanographic endeavors, allows scientists to use the past as a lens for viewing the future, which can ultimately provide evidence-based information to inform policy decisions. The last decadal survey of ocean sciences published for NSF (DSOS-1), titled *Sea Change: 2015-2025 Decadal Survey of Ocean Sciences* (NRC, 2015), realized and emphasized the importance of scientific ocean drilling to the next decade of ocean science research, highlighting five priorities that required at least some component of ocean drilling to be answered (Table 3.1).

In 2023, the White House recognized the urgency of better understanding ocean–climate interactions for informing adaptation and mitigation solutions as a central component of the *Ocean Climate Action Plan* (OCAP). The OCAP priorities are (1) promoting ocean health and stewardship; (2) carbon sequestration in subseafloor geologic formations; (3) supporting ocean research, observations, modeling, and forecasting; and (4) addressing ocean acidification. These can all be informed by addressing the prioritized science areas that require ocean drilling.

Furthermore, the 2022 National Academies report *A Research Strategy for Ocean-based Carbon Dioxide Removal and Sequestration* notes that "very little is known about the timescales of degradation of macroalgal carbon or DNA in seafloor sediments" (NASEM, 2022c, p. 134); addressing these needs falls within the area of future ocean drilling priorities for subseafloor microbiology research. Themes from the consensus study *Cross-Cutting Themes for U.S. Contributions to the UN Ocean Decade* (NASEM, 2022a) are also consistent with the

**TABLE 3.3** Connecting Scientific Ocean Drilling Priorities to U.S. National Priorities and Prior Study Recommendations to NSF

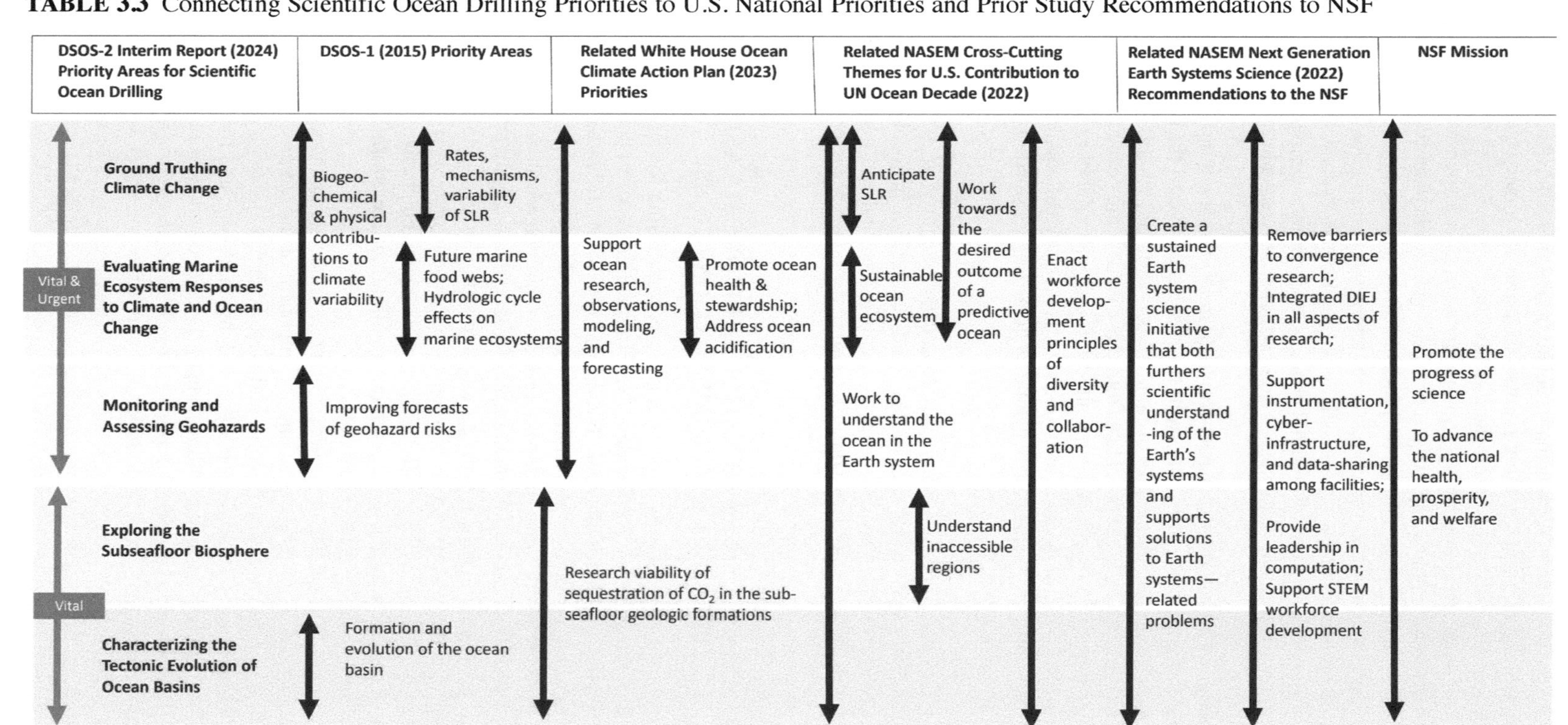

| | DSOS-2 Interim Report (2024) Priority Areas for Scientific Ocean Drilling | DSOS-1 (2015) Priority Areas | Related White House Ocean Climate Action Plan (2023) Priorities | Related NASEM Cross-Cutting Themes for U.S. Contribution to UN Ocean Decade (2022) | Related NASEM Next Generation Earth Systems Science (2022) Recommendations to the NSF | NSF Mission |
| --- | --- | --- | --- | --- | --- | --- |
| Vital & Urgent | Ground Truthing Climate Change | Biogeo-chemical & physical contribu-tions to climate variability; Rates, mechanisms, variability of SLR | Support ocean research, observations, modeling, and forecasting | Anticipate SLR; Work towards the desired outcome of a predictive ocean; Work to understand the ocean in the Earth system; Enact workforce develop-ment principles of diversity and collabor-ation | Create a sustained Earth system science initiative that both furthers scientific understand-ing of the Earth's systems and supports solutions to Earth systems—related problems; Remove barriers to convergence research; Integrated DIEJ in all aspects of research; Support instrumentation, cyber-infrastructure, and data-sharing among facilities; Provide leadership in computation; Support STEM workforce development | Promote the progress of science; To advance the national health, prosperity, and welfare |
| | Evaluating Marine Ecosystem Responses to Climate and Ocean Change | Future marine food webs; Hydrologic cycle effects on marine ecosystems | Promote ocean health & stewardship; Address ocean acidification | Sustainable ocean ecosystem | | |
| | Monitoring and Assessing Geohazards | Improving forecasts of geohazard risks | | | | |
| Vital | Exploring the Subseafloor Biosphere | | Research viability of sequestration of $CO_2$ in the sub-seafloor geologic formations | Understand inaccessible regions | | |
| | Characterizing the Tectonic Evolution of Ocean Basins | Formation and evolution of the ocean basin | | | | |

NOTES: DSOS = Decadal Survey of Ocean Sciences; DEIJ = diversity, equity, inclusion, and justice; NSF = National Science Foundation; SLR = sea level rise; STEM = science, technology, engineering, and mathematics.

guiding principles of scientific ocean drilling and with the science priorities outlined in this report. For example, both include workforce development principles of diversity and collaboration and the desired outcome of a predictive ocean. The priority of *understanding the ocean in the Earth system* aligns with the mission of scientific ocean drilling to conduct global-scale, interdisciplinary research below the seafloor of the world's ocean.

This integrated, systems-based approach to research is also fundamental to the vision for next-generation Earth system science, as laid out by the 2020 NASEM report *Next Generation Earth Systems Science at the National Science Foundation* (NASEM, 2022b) and is central to the *2050 Framework*'s flagship initiatives (see Box 1.2 in Chapter 1) and interrelated strategic objectives (see Figure 1.4 in Chapter 1). Support for facilities and infrastructure toward scientific ocean drilling's mission is consistent with the *Next Generation Earth Systems Science* report's primary recommendation to NSF to "create a sustained next generation Earth system science initiative that both furthers scientific understanding of the Earth's systems and supports solutions to Earth systems–related problems" (NASEM, 2022b, p. 6). The report further recommends that NSF remove barriers to convergent research and support a range of instrumentation and data initiatives, as well as diverse workforce development; these are also identified as enabling elements to priority science in the *2050 Science Framework* (Koppers and Coggon, 2020) and are supported in this consensus study.

**CONCLUSION 3.5** The vital and urgent scientific ocean drilling research priorities connect and respond to U.S. research priorities identified by the White House, by the scientific ocean drilling community, and by several National Academies studies.

**CONCLUSION 3.6** The rapid pace of climate change and related extreme events, sea level rise, changes in ocean currents and chemistry impacting ocean ecosystems, and devastating natural hazards such as earthquakes are among the greatest challenges facing society. Scientific ocean drilling research continues to play unique and essential roles in addressing these challenges.

# 4

# Needs for Accomplishing the Science Priorities

The history of science is replete with examples of how scientific curiosity and societal need have driven technological advancement, as well as converse examples where innovative technological advancements have spurred scientific inquiry. Throughout all fields of science—ocean sciences are no exception—the swinging pendulum between scientific discoveries and technological innovation has resulted in robust, decadal-scale (or longer) paradigm shifts, significant advancements in knowledge, and economic growth spurred by the joining hands of science and technology.

Scientific ocean drilling in the United States has been dominated by a single platform, from the very beginning of the Deep Sea Drilling Project (DSDP) (with the Glomar *Challenger*), through the Ocean Drilling Program (ODP), the Integrated Ocean Drilling Program (IODP-1), and the International Ocean Discovery Program (IODP-2) (for the latter three programs, the *JOIDES Resolution* has been the primary platform). Although significant discoveries and technological advancements have been made with the *Chikyu* and mission-specific platforms (MSPs), as articulated in this report, the scientific needs of the U.S. and international communities have been most ably served by a single globally ranging platform over a period of decades. Significant technological and analytical innovations have been made throughout these decades, speaking to the "swinging pendulum" model of advancement.

These observations raise the question of whether the committee's present perspective on vital and urgent priorities is constrained by the scientific ocean drilling program's long history of using a single platform. Alternatively, is the committee's view of, and ability to address, these vital and urgent challenges a result of having worked with and developed a platform uniquely suited to addressing these questions?

Chapter 4 examines the science priority areas identified in Chapter 3 through the lens of those components that can be achieved with currently archived materials (e.g., sediment and rock cores, logging and borehole datasets) and other recovery methods (other existing drilling platforms). Furthermore, the chapter describes how the accomplishments of past drilling programs can best be utilized for domestic and international collaboration, and how to ensure effective use of science to develop solutions for the grand challenges that Earth faces, such as resource depletion, the changing climate, and escalating hazards.

## REALITIES OF EXISTING AND EMERGING DRILLING TECHNOLOGIES

In 1859, the first crude oil well was drilled in Titusville, Pennsylvania, thereby establishing an entire new, paradigm-shifting industry. Originally driven by economic demand for oil and gas resources, technologies regarding drilling, and to a lesser extent coring, have been evolving for more than 150 years. Although it was initially land based, the commercial industry has made continual technological and engineering forays into offshore drilling and into progressively deeper (yet nearshore) waters, as demand for cheap and plentiful energy resources has grown.

Scientific drilling is motivated by the pursuit of knowledge and has a far younger history than commercial drilling, beginning only over the past 50 years, as described earlier in this report. Like drilling for energy, scientific drilling has historically been led by the United States. Scientific drilling has benefited from engineering skills and expertise from the oil and gas industry, including the capability to drill at sea with drillships. Industry, in turn, has benefited from the scientific and technological innovations developed via scientific ocean drilling. This mutually beneficial association has opened new horizons for scientific progress. A key critical and insurmountable difference, however, is that scientific ocean drilling requires recovery of sediment and rock cores, whereas industry-based drilling rarely cores. Because of this, the technologies have necessarily diverged.

Existing and emergent technologies can be deployed in pursuit of the five Decadal Survey of Ocean Sciences (DSOS-2) priority research areas identified in Chapter 3 for future scientific ocean drilling. Identifying these technologies requires a matrix of considerations, given the variety of environments and geographies in which scientific ocean drilling is conducted. For example, scientific drilling in very shallow waters (10–100 m) can be achieved through mobile offshore drilling units, such as "jack-up rigs" or semisubmersible platforms, which are commercial industry–based technologies that can be modified for scientific drilling. Deep-water scientific drilling, however, (loosely meaning deeper than 100 m of water, but most commonly meaning many thousands of meters), requires larger ocean-capable vessels equipped with dynamic positioning (to position the vessel in a fixed location above the ocean floor) and thousands of meters of drill pipe available for immediate use on the vessel. Of high relevance to the profound societal challenges regarding climate change, scientific drilling in areas where ice may be encountered and severe weather conditions at high latitudes is routine, requiring unique technologies and expertise. While MSPs have been deployed successfully for some of these shallow-water and/or high-latitude objectives, relatively few MSPs have been deployed throughout the history of scientific drilling (Table 1.1), and only one MSP associated with scientific ocean drilling has operated in the Arctic.

For the United States, the ability to collect long and deep cores by drilling is restricted exclusively to the scientific ocean drilling program. In contrast, the Academic Research Fleet (ARF) overseen by the University-National Oceanographic Laboratory System (UNOLS) can be used to recover sediment from shallower subseafloor depths. Shorter (penetrating to shallow subseafloor depths) devices can be deployed from a range of UNOLS vessels, whereas long piston cores (20–50 m) require larger vessels with the ability to deploy heavy tools (A-frames, winches, and associated engineering requirements, including dynamic positioning). With the retirement of the *JOIDES Resolution*, there will be no means to piston core deeper than 50 m (at best). Furthermore, the ARF currently has no vessel in its fleet large enough to support deployment of long or giant piston corers that can recover up to 50-m cores (and has no plans to add one).

Less than 5 percent of current scientific ocean drilling objectives would be achievable with the current U.S. vessels of the ARF (which are supported by the National Science Foundation [NSF]) (Figure 4.1). Even restoration of 50-m+ giant piston core capabilities would enable only up to 10 percent of the ocean drilling science objectives. Additionally, to the committee's knowledge, no privately funded vessel is able to meet these deep-subseafloor objectives. Therefore, **at least 90 percent of scientific ocean drilling objectives will not be met if the United States is dependent only on the post-2024 ARF**.

Remotely operated seabed lander–based drilling systems, a technology emerging over the past decade that potentially opens new opportunities for scientific ocean drilling, can recover up to 260 m of subseafloor sediment using a multihole operational approach in certain settings and may offer a partial solution to achieving goals of the scientific ocean drilling program. However, the capability to operate these systems would need to be developed for the U.S. ARF, as the current fleet does not have this capability, and such systems cannot host deep observatories (>100 m below seafloor) or perform significant downhole logging. Because of these limitations, whether or not lander-based drilling was available, **any scientific objectives that require drilling deeper than 260 m beneath the seafloor could be achieved only using a drilling vessel**. There is currently no way to sample those depths with any other configuration of current technology (including modifying existing technology or partnering with industry assets).

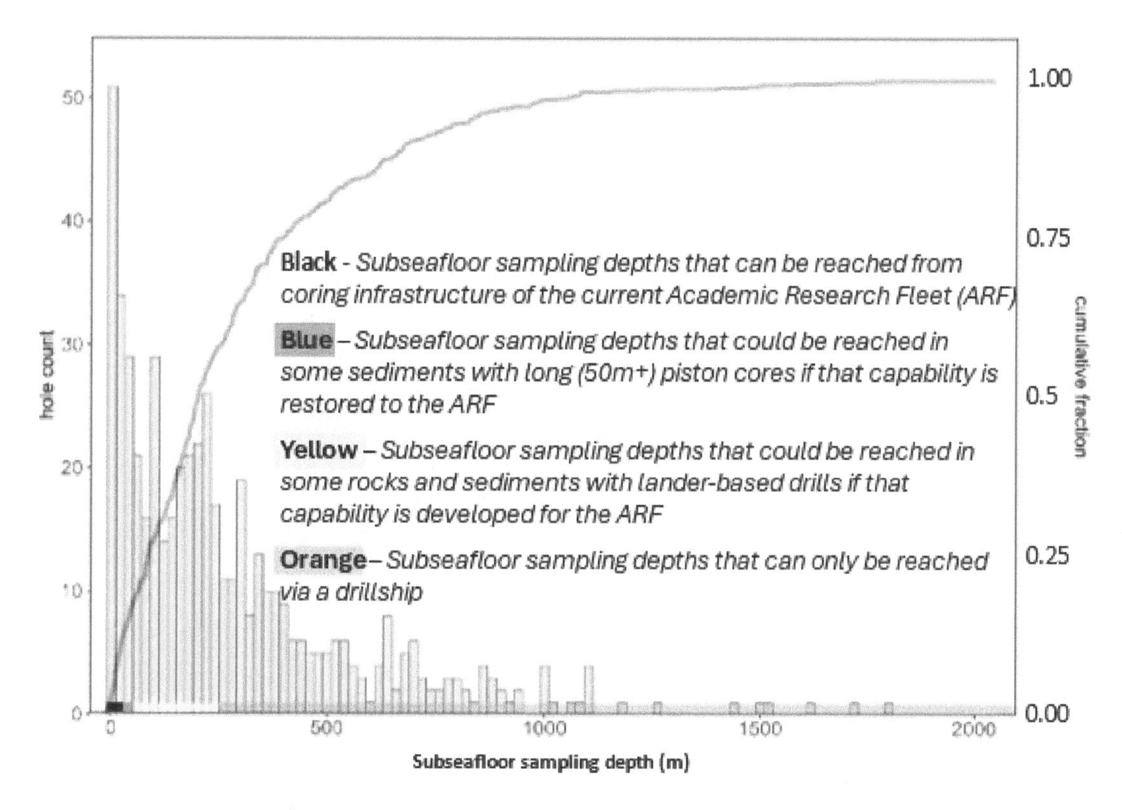


**FIGURE 4.1** Capabilities of the current U.S. Academic Research Fleet (ARF) for subseafloor sampling. NOTES: The current ARF, representing vessels of the University-National Oceanographic Laboratory System, has many opportunities for recovering material at the sediment–water interface, seafloor rock exposures, and cores that are <10 m long. However, it currently has only one option for recovering cores between 10 and 30 m long, and no options for recovering cores >30 m long. The red curve represents the cumulative fraction of the holes at different seafloor penetrations. The blue horizontal bar demonstrates that nearly 90% of scientific ocean drilling objectives cannot be met if the United States is dependent on only the post-2024 ARF. As shown with the yellow horizontal bar, remotely operated seabed lander–based drilling systems, which can reach up to 260 m subseafloor, would offer a partial solution if such capability is developed for the ARF. Any scientific objectives below 260 m subseafloor will require a drilling vessel, as there is no way to sample those depths with any other configuration of current technology.
SOURCE: Maureen Walczak, Oregon State University; data courtesy of Maureen Walczak and Laurel Childress (Texas A&M University).

**Today, neither long (50- to 60-m) giant piston coring nor lander-based drilling can be deployed from a vessel in the ARF; some combination of investment in the infrastructure and the capabilities of the vessels themselves would be required to use coring technology.** Both technologies are available "off the shelf" and can be deployed from sufficiently large, general-purpose oceanographic research vessels, as well as commercial platforms such as "mud boats," with required modifications to those vessels. However, the modifications are likely to be of the scale that would cost tens of millions of dollars and, under ideal circumstances, would recover only soft materials from roughly 50 to 250 m below the seafloor (50 m for giant piston cores or roughly 250 m for lander-based recovery), which would render many goals of the scientific priority areas unachievable. Additionally, hole diameters for at least some lander-based drilling systems can be smaller than standard holes drilled by the *JOIDES Resolution,* which could impact core recovery and the ability to install subseafloor observatories.

Thus, with the decommissioning of the U.S.-funded drilling vessel, most of the global ocean subseafloor will no longer be accessible using the existing U.S. ARF (Figure 4.2). In addition, deep hard-rock cores, borehole

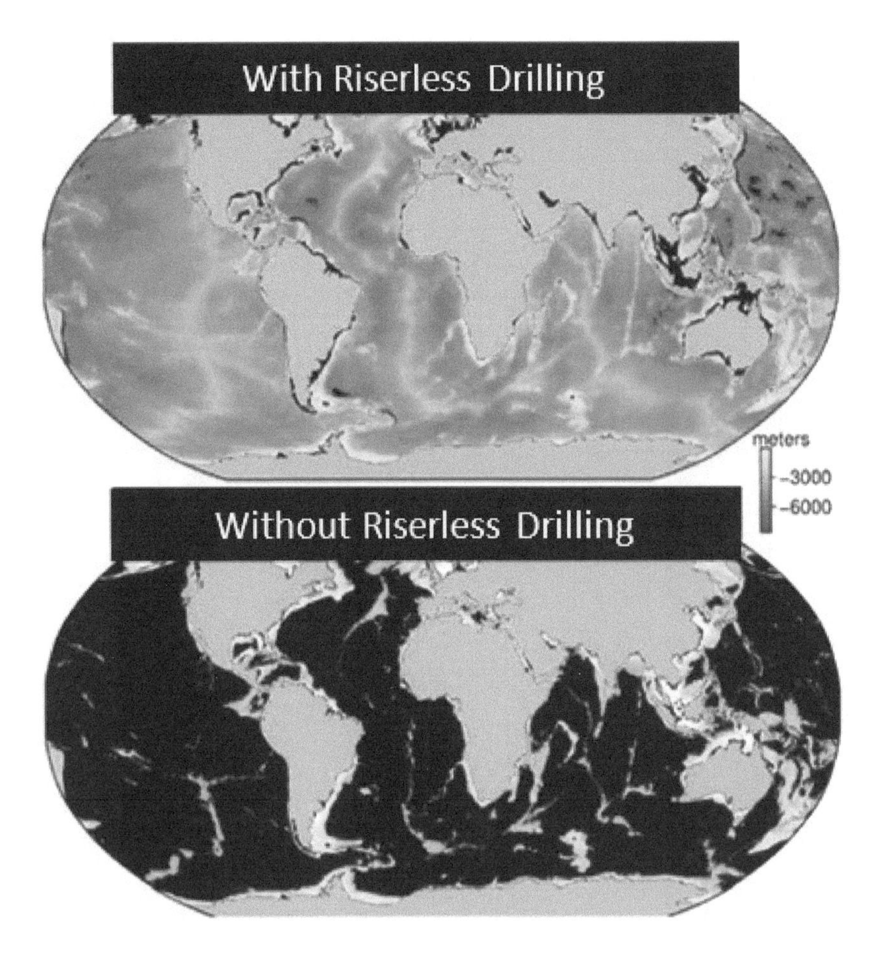


**FIGURE 4.2** The impact of riserless drilling in access to global subseafloor depths. NOTES: Black regions in both global ocean maps indicate the regions that will be inaccessible for scientific ocean drilling without a drilling technique that captures sediments past 260 m of sediment, the deepest penetration of seafloor sediments that can be obtained outside of riserless drilling technologies (see Box 3.1 in Chapter 3). With the decommissioning of the U.S.-funded *JOIDES Resolution* in 2024, most of the global ocean subseafloor records will no longer be accessible using the existing U.S. Academic Research Fleet. In addition, deep hard-rock cores and borehole observatories are unlikely to be completed.
SOURCE: Ross Parnell-Turner and Anthony Koppers.

observatories, and a host of other measurements would not be completed. In this context, it is important to recall that the *JOIDES Resolution* could be operated in some capacity until 2028, when its permit (Environmental Impact Statement) expires. A gap in drilling capability will impede achievement of multiple science goals (Chapter 3). It will also impact the level of expertise retained in the workforce, across career stages and disciplines, requiring decades to rebuild. **To the extent that it remains a viable option operationally and financially, the *JOIDES Resolution* could serve as an MSP in the near term.**

## OPPORTUNITIES FOR USING EXISTING CORES, BOREHOLES, AND DATA

Opportunities exist for using available assets, including collected cores, data, and other samples, to continue to accomplish groundbreaking and essential scientific research. However, these opportunities will not replace the need for drilling technology to collect new cores and develop new observatories from an operating vessel.

A holistic approach to understanding the scientific priority areas outlined in Chapter 3 includes strategic use of existing archives along with targeted drilling for new records.

## Existing Data, Core Archives, and Borehole Observatories

In line with current NSF policy, all data collected as part of the scientific drilling program are shared openly. **Sharing of data is critical for the community to continue to thrive and learn from what has already been collected**, especially if there is a time gap between when new cores are collected and/or observatories are installed or occupied. While issues arise with using existing cores, such as deterioration of quality core material, **there is value in using existing cores, supporting data, and other samples for further study and analyses**. In fact, given adequate funding (primarily but not exclusively from NSF), the scientific ocean drilling community can expand the use of existing materials.

Available assets and their limitations include the following:

- **Cores:** Approximately 150 km of collected core from all drilling platforms are stored in each of three locations: Gulf Coast (United States), Bremen (Germany), and Kochi (Japan), for a total of around 450 km. Approximately one-third of the total core length is appropriate for high-priority science. Cores appropriate for high-resolution paleoceanographic studies are critical for understanding the dynamics of rapid climate transitions and feedback. However, such cores, especially those of high scientific and societal interest, are quickly depleted by use (e.g., Figure 4.3). Additionally, cores taken by the legacy programs DSDP, ODP, and earlier expeditions of IODP-1 are now dried, and some are contaminated by mold (a natural consequence of long-term storage), which makes obtaining chemical data from these older sediments challenging.
- **Microbiology samples:** Approximately 1,300 samples exist that are frozen for preservation for molecular analysis. However, freezing commonly limits usability in future analysis, and past experience has highlighted challenges in storage. For example, frozen samples are not suitable for determining microbial activity or rates of activity and are unsuitable for any potential cultivation work. While many biological analyses can be carried out using frozen materials, neither physiological nor direct metabolic studies can be carried out with frozen materials. Such efforts require fresh material.
- **Data** (measurements, imagery, and metadata): Data include collecting standardized measurements of cored material and those collected via logging the drilling holes. There are approximately 1,000,000 unique measurements per drilling expedition, and around 700 core images and 700 X-ray images per kilometer of each existing core. There will always be issues with data quality, resolution of the time record, calibration, and combining datasets. Continued support for data stewardship activities is critical to handle issues related to data quality, resolution of the time record, calibration, and combining datasets.

**FIGURE 4.3** Highly sampled core from an interval of high scientific interest. NOTES: Styrofoam spacers (white) replace samples that have already been used. ~40-cm interval of Paleocene–Eocene thermal maximum (PETM) at Site 1215, Leg 199, 1,000 km NE of Hawaii. Interval is ~55 m into the seafloor. (The PETM was a time of extreme warmth and high carbon dioxide within the range expected by the year 2100.)
SOURCE: *JOIDES Resolution* Science Operator.

- **Instrumented boreholes:** Instrumented boreholes provide 10–50 times the sensitivity of bottom-mounted sensors for many critical measurements and are highly valuable assets that can be used for key studies in certain areas of oceanography beyond geological interests (e.g., geochemistry, microbiology, hydrogeology). Approximately 50 active borehole observatories exist, but few transmit data in real time, and they require revisitation to install/reinstall apparatus and/or download data. Additionally, about 90 inactive borehole observatories are ready for reentry and reinstrumentation by a vessel, if determined practical and feasible.

## What Can Be Achieved With Existing Data, Core Archives, and Borehole Observatories

*Legacy Asset Projects*

A new approach to collaborative research has been proposed by the scientific ocean drilling community, with the first call for proposals for Legacy Asset Projects (LEAPs) issued in October 2023 by the IODP Science Support Office. LEAPs encourage exploration of archived cores and existing samples, without requiring new drilling. The objective of the proposed LEAPs program is to maximize the scientific value of legacy assets already collected from scientific ocean drilling while addressing the *2050 Science Framework* (Koppers and Coggon, 2020). LEAPs could be thought of as "virtual expeditions," and they also enable big data analytics. However, it is important to note that at the present time, no funding is dedicated to support proposed LEAPs projects.

Providing a specific funding call to analyze legacy data and cores is an opportunity for NSF to support projects that conduct multidisciplinary integration, synthesize existing data, and explore data in new ways. It has the potential to stimulate future drilling expeditions and serve as an incubator for new ideas. Funding LEAPs will encourage involvement and participation from a broad community with potential to involve minoritized and/or historically excluded identities, leading to diverse science parties and opportunities for early-career scientists. LEAPs support may also increase visibility or reenhance outcomes of existing projects and the overall scientific contribution of scientific ocean drilling. **If funded, LEAPs provide a new and flexible mechanism for large, multidisciplinary, community-driven research efforts, maximizing the return on legacy assets of past scientific ocean drilling and strengthening the impacts of past funded research**.

However, the science that can be done through LEAPs is not a replacement for recovering new core and sample assets in the future, given the state of the assets that exist today. LEAPs suffer several limitations, two of which are the **inability of stored material to address many science questions of urgent and vital high-priority areas**, as identified in Chapter 3, and that some of **the most important core materials have been (appropriately) depleted by use** and yet remain in high demand (thus new material is required).

**CONCLUSION 4.1** While a funded LEAPs initiative can augment drilling for newly recovered material, it is not a long-term replacement of drilling capability. LEAPs provide an opportunity to maximize the use of already acquired material and data and foster discovery and innovation. An ideal scientific drilling program could include a robust LEAPs program combined with recovery of new subseafloor cores and installation of borehole observatories that address the five high-priority research areas.

Table 4.1 summarizes what can and what cannot be done using existing assets, organized by the five high-priority areas identified by this committee.

*Open Data Sharing*

The universe of data associated with scientific ocean drilling includes data gathered onboard the drilling platform during an expedition, data published in the scientific literature resulting from research conducted after an expedition (and sometimes decades later), as well as borehole data. Even though the data are considered broadly similar, they vary in formats because they were collected using past technology. While funding use of legacy assets is important, it is equally important that the **metadata and data collected, regardless of type or source, be findable, accessible, interoperable, reproducible (FAIR), and shared in a timely manner**.

**TABLE 4.1** High-Priority Science Areas That Can and Cannot Be Addressed Using Existing Assets

| Priority Areas for Future Scientific Ocean Drilling in This Report (DSOS-2) and Their Available Legacy Assets | What Science Areas *Can* Be Addressed with Legacy Assets? | What Science Areas *Cannot* Be Addressed (or Are Significantly Limited) with Legacy Assets? |
| --- | --- | --- |
| **General/Cross-Areas** | Community-driven, collaborative, multidisciplinary research.<br>Big data analytics on a wide range of subseafloor standard measurements (e.g., physical properties, petrophysics/logging, paleomagnetic data).<br>Large-scale "syntheses of science" studies (i.e., producing topical review papers) that integrate data across multiple expeditions/boreholes, addressing global or regional geographies and time intervals.<br>Development and testing of new proxy methods (that are not dependent on ephemeral properties).<br>Undergraduate and graduate education and training on materials and methods used in scientific ocean drilling research to help keep early-career scientists engaged with scientific ocean drilling. | High-resolution, sample-intensive studies for cores that have already been heavily sampled (e.g., many cores include recovered records of the Paleocene–Eocene thermal maximum [PETM], a very high $CO_2$ world, but the intervals of primary interest [see Figure 4.3] are already heavily sampled).<br>Comprehensive studies of igneous and metamorphic rocks would not be possible because very little repository material of these rock types is available.<br>Studies of challenging rock types, such as those in fault zones, because little repository material of such intervals is available from locations other than those along the Japan margin.<br>Real-time monitoring of fault motion using existing borehole instruments.<br>Microbiology studies on living microbes.<br>Analyses that depend on ephemeral properties (e.g., pore water, organic carbon).<br>Coordinated land–sea studies, because existing assets were not necessarily taken from locations best positioned for linking to adjacent continental records (which may, themselves, not be available yet). |
| **Ground Truthing Climate Change**<br><br>**Legacy assets available:**<br>400–500 km of core, but intervals of high scientific and societal interest (e.g., transient climate states) are a small portion of the total core holdings, have been poorly recovered in existing cores, or already have been sampled extensively. Some geographic gaps (e.g., equatorial transects, midlatitude transects, only one drilling site in the Arctic Ocean). | Improve understanding of regional patterns of climate change, particularly during warmer intervals, if using proxies that do not degrade after cores are taken.<br>Coordinate with Earth system modelers to provide ground truth data on possible directions and magnitudes of system feedbacks.<br>Develop and apply new age-dating techniques.<br>Continue to apply multiproxy approaches across a range of temporal and spatial scales.<br>Conduct/support research specifically designed to synthesize existing but separate proxy datasets into multiproxy datasets. | Evaluate relationships between ice sheet extent and various aspects of the global carbon reservoirs and carbon cycle, necessary to inform climate models.<br>Identify new climate "tipping points" that requires synchronous, high-temporal-resolution records from multiple climatically/oceanographically sensitive regions.<br>Identify decadal- to millennial-scale regional climate variability, which document lead/lag relationships on high-resolution timescales useful for predictive models.<br>Track climatic changes and their effects (e.g., aridity, seasonality of precipitation) from land to sea by coordinating studies in both regions, limiting research progress on societally relevant regional scales.<br>Apply any paleoclimatic proxies that use materials subject to degradation after cores are taken (e.g., organic-based proxies).<br>Address emerging questions regarding rates of future climate change in the mid- or high latitudes, and in ocean gateways. |

*continued*

**TABLE 4.1** Continued

| **Priority Areas for Future Scientific Ocean Drilling in This Report (DSOS-2) and Their Available Legacy Assets** | **What Science Areas *Can* Be Addressed with Legacy Assets?** | **What Science Areas *Cannot* Be Addressed (or Are Significantly Limited) with Legacy Assets?** |
|---|---|---|
| **Evaluating Marine Ecosystem Responses to Climate and Ocean Change**<br><br>**Legacy assets available:** 400-500 km of core, but intervals of high scientific and societal interest (e.g., responses to transient climate states, ocean acidification, biodiversity stressors and natural experiments in changing productivity/oxygenation/nutrient supply) are a small portion of the total core holdings, have been poorly recovered in existing cores, have already have been sampled extensively, or have been affected by chemical reactions during core storage which degrade the integrity of carbonate fossils and molecular fossils (biomarkers). Some geographic gaps (e.g., equatorial and midlatitude transects; only 1 drilling site in the Arctic Ocean). | Conduct/support research specifically designed to integrate data across different paleoecology fossil groups (e.g., molecular fossils, carbonate and siliceous mineralized fossils, organic-walled marine and terrestrial fossils).<br>Develop/expand species-level databases of Cenozoic planktic and benthic fossil occurrences useful for macroecological and macroevolutionary studies on species' responses to climate and ocean change.<br>Improve understanding of global and (some) regional ecosystem responses to climate and ocean changes.<br>Examine land–sea changes in ecosystems from palynomorph fossils (e.g., terrestrial pollen and spores) transported and preserved in ocean sediment cores. | Conduct studies of long-term ecosystem responses from marginal seas to open ocean settings where drilling cores (e.g., near- to offshore site tracks; longitudinal site tracks) do not exist, resulting in contextual gaps when evaluating ecosystem variability.<br>Design new studies of ecosystem responses at decadal to millennial scales, which would require high-temporal-resolution records that are not available in the legacy assets.<br>Address new questions of ecosystem responses in equatorial, some midlatitude, and high-latitude regions (especially the Arctic) because there are few records from these settings, resulting in contextual gaps when evaluating ecosystem variability.<br>Studies involving carbonate microfossils and molecular fossils given reduced utility (chemical alterations) of older cores (those in storage longer). |
| **Monitoring and Assessing Geohazards**<br><br>**Legacy assets available:** existing cores, borehole observatories, and datasets. However, intervals of high scientific interest may have limited availability, either because they are in the form of lithologies that are difficult to core and recover or because they have already been sampled extensively. Most existing observatories would not be considered state of the art, and only a handful are monitored at anything approaching real time. Geographic distribution of observatories is limited. | Provide evidence of the magnitudes and recurrence intervals of geologic hazards in areas for which recovered materials (e.g., intervals of fault slip or transported material, and explosive eruption products) exist.<br>Monitor existing observatories for seafloor geodesy. | Significantly improve understanding of precursor events to geologic hazards (e.g., explosive volcanic eruptions, fast-slip vs. slow-slip earthquakes, tsunami-generating events), impacting efforts toward hazard predictions.<br>Improve dynamic models of these hazards without additional data from in situ monitoring capabilities.<br>Monitor seafloor geodesy (underwater Earth surface deformation and displacement). Observatories necessary for time-dependent hazard assessment do not currently exist but are urgent and relevant in places such as the Pacific Northwest (e.g., Cascadia).<br>Achieve rapid-response capabilities in the case of large subduction-zone earthquakes and tsunamis, such as those threatening Cascadia.<br>Extend records of submarine volcanic eruptions and landslides back in time and improve constraints on slope failure hazards by sampling in situ materials from these zones. |

*continued*

**TABLE 4.1** Continued

| **Priority Areas for Future Scientific Ocean Drilling in This Report (DSOS-2) and Their Available Legacy Assets** | **What Science Areas *Can* Be Addressed with Legacy Assets?** | **What Science Areas *Cannot* Be Addressed (or Are Significantly Limited) with Legacy Assets?** |
|---|---|---|
| **Exploring the Subseafloor Biosphere**<br><br>**Legacy assets available:**<br>~1,300 samples, stored frozen. | Determine what microbes are present in the sampled intervals, although some biological materials are damaged by freezing. | Quantify the microbial ecosystem and identify new species.<br>Conduct more comprehensive spatial and temporal surveys of the seafloor microbiota. Limited to materials on-hand and because this is a newly emerging field, there are large gaps in data.<br>Assess and refine new techniques for sampling, sample storage, and analyses.<br>Assess the chemical reactions that the microbiota were conducting, or their reaction rates (including metabolic activity). This limits the ability to better understand the role of subseafloor microbiota in important global geochemical cycles.<br>Evaluate limits of microbial life (e.g., temperature, pressure, salinity, age).<br>Assess the role of subseafloor microbial life in the global carbon budget, and its associated feedbacks. |
| **Characterizing the Tectonic Evolution of Ocean Basins**<br><br>**Legacy assets available:**<br><45 km of crustal core; estimated <100 holes have been drilled that penetrate >100 m into basement. For examining ridge processes, legacy assets are dominated by materials from seven holes. Approximately 50 instrumented boreholes (i.e., Circulation Obviation Retrofit Kits [CORKS] and their successors), with only a handful monitored. | Extend understanding of ocean crust formation at slow-spreading centers, ocean crust maturation and hydrothermal circulation, and hotspot origin and evolution.<br>Understand the role of seafloor crustal processes as natural $CO_2$ sinks. | Independently assess models for fast-spreading and slow-spreading ridges when only a few of each have been adequately drilled/sampled because of drilling challenges (e.g., scientists cannot use the same data to test a model that were used to develop that model).<br>Develop spatial and temporal large-scale records of rock alteration that could inform the impact of these processes on the past global carbon budgets.<br>Assess the ocean's potential for capture, removal, and burial of modern $CO_2$ in ocean crust to help mitigate anthropogenic warming.<br>Understand spatial and temporal variability in circulation through, and processes occurring within, the seafloor aquifer, because few observatories are monitored. |

SOURCES: Informed in part by Larry Krissek (*JOIDES Resolution* Facility Board Chair, personal communication), the Legacy Asset Projects (LEAPs) working group report, and LEAPs proposal guidelines (see https://www.iodp.org/call-for-leap-proposals).

In order to make data sharing a reality, the following actions would be necessary: (a) ensure that the legacy shipboard data remain accessible and comply with the FAIR principles, (b) empower researchers generating postcruise data to improve their data stewardship and management, and (c) catalog and integrate borehole data into the rest of drilling science. For example, the Extending Ocean Drilling Pursuits (eODP) Project (Sessa et al., 2023) has established a workflow for compiling, cleaning, and standardizing scientific ocean drilling records and importing them into existing open-access database systems (e.g., Paleobiology Database, Macrostrat) (Figure 4.4). At the time of publication, eODP had processed all of the lithological, chronological, and paleobiological data from one scientific ocean drilling repository; the compiled dataset contains nearly 80,000 lithological units from 1,125 drilled holes from 422 sites.

Integrating scientific ocean drilling data into existing open-access database systems is an important step toward data management and data sharing. However, additional strategies to further enable more widespread use of the data may also be needed. Partnering with big data experts (perhaps via funding through NSF's Directorate for Technology, Innovation and Partnerships), may enable the data to be more impactful, if it is shared correctly with use-inspired intent. However, use-inspired sharing will not mitigate issues with data quality, calibrations, and integrating datasets. **Specific funding opportunities are needed to support data stewardship initiatives.** Paired with meaningful and robust accountability and oversight, great gains in data science could be realized by the scientific ocean drilling community.

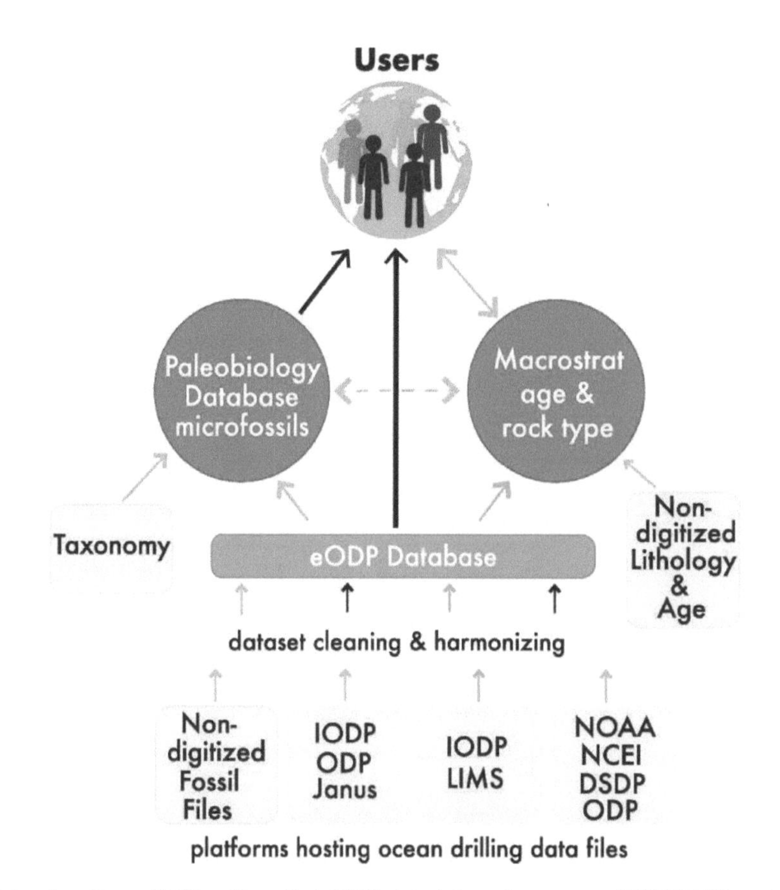


**FIGURE 4.4** The Extending Ocean Drilling Pursuits (eODP) data-integration system. NOTES: The schematic representation displays data sources, data types, databases, and connectivity among the various components. eODP is focusing on three data types: age models, fossil occurrence and abundance data, and lithology. IODP = International Ocean Discovery Program; LIMS = Laboratory Information Management System; NCEI = National Centers for Environmental Information; NOAA = National Oceanic and Atmospheric Administration.
SOURCE: Figure and caption from Sessa et al., 2023.

Furthermore, **timely sharing of all newly collected data is essential and could be incentivized and valued to the same extent as publications of expedition outcomes.** If funding agencies such as NSF establish data stewardship models that acknowledge use-inspired data sharing as valuable as a scientific publication, the culture would shift to one of FAIR and timely data sharing, not only in the scientific ocean drilling program, but throughout the field of ocean science.

### Communicating Worth

Ocean drilling research has sparked new scientific and technical knowledge, leading to a better understanding of Earth's past and present events and systems. The excitement generated by scientific ocean drilling has attracted students from different backgrounds and nationalities to careers in science, technology, engineering, and mathematics; provided them with research opportunities; fostered mentorship; and contributed to the ocean science workforce capacity. It has facilitated interdisciplinary training in Earth and life sciences and marine engineering and technology. Scientific ocean drilling has provided the content for science textbooks, thereby influencing multiple generations of students and future contributors to society. Education and outreach programming aboard vessels, such as the School of Rock program (IODP, n.d.h), have provided professional development opportunities for college and university faculty and K–12 teachers to learn not only what is known about the Earth and ocean system, but how it is known.

While the ocean drilling community has had a clear record of scientific success and transformative impacts on cohorts of educators, students, and early-career researchers, it has fallen short in communicating its worth in other spheres. In general, scientific ocean drilling is largely unknown in the U.S. public and policy realms. The U.S. scientific ocean drilling community is aware of these shortcomings and challenges, based on recent workshop discussions and findings (Chapter 1). While each IODP-2 expedition included outreach components, these were essentially encapsulated one-off endeavors. With such an approach, it is difficult to see how stepwise advances in public awareness of the scientific and societal value of the program can be made. **Excellent science communication is an essential component to any large-scale, long-term research endeavor; programwide coordinated, strategic science communication and branding are important to the future of scientific ocean drilling.**

In addition to communicating with the public and policy makers, it is also important to communicate the worth of scientific ocean drilling to scientific communities with overlapping research interests and goals. The five high-priority science areas identified in this report intersect with national science and technology priorities and recommendations (see Table 3.3 in Chapter 3). Broadening the sphere of scientific ocean drilling expedition and/or LEAPs proposal writing teams to involve, for example, biological and chemical oceanographers, and climate and ecosystem modelers may strengthen the science and its impact. **Interdisciplinary networks and collaborations across scientists and engineers from multiple disciplines are required to address vital and urgent scientific ocean drilling research priorities identified in this report.**

## ADDITIONAL INFRASTRUCTURE NEEDS

Based on the five high-priority areas for future scientific ocean drilling (Chapter 3), the committee identified key criteria, or parameters, for successful achievement of the scientific goals associated with each area. Rather than recommend to NSF any specific path forward in terms of infrastructure, the committee aimed to identify parameters necessary for successful fulfillment of the vital and urgent scientific themes.

These parameters, listed as the column headings in Table 4.2, are not intended to be exhaustive. Instead, they define the most high-level screening parameters relevant to generating the data scientists need to address the themes, which are listed as rows in Table 4.2. This table does not capture other critically important components of scientific ocean drilling, such as workforce needs or other topics described later in this section.

> **CONCLUSION 4.2** While some scientific research objectives can be accomplished using existing assets, many science objectives critical to U.S. interests cannot be accomplished without new and/or continuing scientific ocean drilling assets.

**TABLE 4.2** Parameters for Accomplishing Vital and Urgent Ocean Drilling Science Research Priorities

| | **Deep Water** >3,000 m water | **Deep Penetration** >30 m sediment/ rock | **Continuous Records from Cores** No unknown gaps in recovery | **Ephemeral Properties** Porewater, Magnetics | **Borehole Observatory/ Instrumentation** Chemistry, Physics, Biology, Geology | **Logging** Downhole tools after coring | **Ice Strengthened (not icebreaker)** |
|---|---|---|---|---|---|---|---|
| **Ground Truthing Climate Change** | **R** | **R** | **R** | **R** | **NN** | **G/NN** | **R** |
| | Records are from all world's ocean environments | Required for old records AND younger records with high resolution (high sed rates) | Requires multiple recoveries per location with intentional offsets | To document potential alteration of physically recovered material | | Cannot replace continuous records from cores | Dependent on target of interest |
| **Evaluating Marine Ecosystem Responses to Climate and Ocean Change** | **R** | **R** | **G** | **G** | **G/NN** | **NN** | **R** |
| | Records are from all world's ocean environments | Required for deep-time biotic events | Required to infer timing and tempo of ecological response to environmental perturbations | Chemical fluxes upward from the seafloor to the ocean are indicators of and sustain deep life | | | Dependent on target of interest |
| **Monitoring and Assessing Geohazards** | **R** | **R** | **R/I** | **R/I** | **R** | **I/R** | **G** |
| | Continental margin and trenches, deep-water records of volcanic ash, midocean ridge relationships | Deep seismogenic zones, old records of recurrence | Dependent on target of interest (yes to temporal earthquake records, eruption records), multiple recovery not required | Dependent on target of interest | Very strong requirement for time-dependent hazards assessment | Perhaps could replace continuous core recovery in certain cases | Dependent on target of interest |
| **Exploring the Subseafloor Biosphere** | **R** | **R** | **G/NN** | **R** | **I** | **NN** | **G/NN** |
| | Organic matter supply varies with depth and distance from shore | Habitability in low-energy substrates | Depth and age (only) necessary | Very strong requirement | Dependent on target of interest | | Dependent on target of interest |
| **Characterizing the Tectonic Evolution of Ocean Basins** | **R** | **R** | **R/I** | **G/NN** | **R** | **R** | **I/G** |
| | Midocean ridges and old oceanic crust are deep water | Establishing crustal boundaries and accessing crust beneath buried sediments | Depending on target of interest (see Logging) | | Very strong requirement | Due to challenging recovery of hard rock | |

NOTES: G = good if a byproduct of a primary driver; I = important, but not required by itself; NN = not necessary; R = required.

## Workforce Needs

The need for a diverse, equitable, and inclusive workforce in ocean science and engineering has been noted in Chapter 1 and is reiterated here as an infrastructure component fundamental to the advancement and future success of scientific ocean drilling. With that underpinning, **a trained workforce skilled in the planning, collection, analysis, and archiving of scientific samples and data has been and will continue to be critical to the future of ocean sciences in its entirety; ocean drilling contributes significantly to this goal.**

Scientific ocean drilling is in many ways a seedbed for successive generations of ocean scientists. Currently there is generally gender equity in expedition science teams, and early-career scientists and graduate students typically comprise two-thirds of expedition science team members (see Box 2.2 in Chapter 2). Decades of attention to such demographics are yielding positive results, although more progress is needed.

The scientific ocean drilling community has played a leadership role in producing scientists who use an integrated Earth system–based framework. Many of these individuals will have careers in ocean sciences and beyond, may move to industry or government positions, and are tuned to international collaboration. Scientific ocean drilling also employs a highly skilled technological and engineering workforce. As mentioned earlier, the nature of commercial industry–led drilling is quite different from scientific drilling. Scientific drilling to obtain intact cores is extremely difficult, and the capabilities of the current workforce have been developed through generational handoffs of know-how and techniques, which cannot be gleaned from a textbook. A challenge in the post-2024 landscape of scientific ocean drilling will be retaining and/or training the technical workforce to support future drilling efforts.

## Management and Staffing Infrastructure

The management and staffing structure for the current IODP-2 program is designed for multiplatform expeditions, a complex international constituency, a global reach, multigovernment funding, and other aspects of an all-encompassing program. It includes core repositories on three continents. A nimble and focused management structure is key to a sustainable and successful future U.S.-based scientific ocean drilling effort. Management and staffing requirements will depend on the nature of the U.S. program design—for example, whether the program uses a U.S. MSP or centers on an acquired (through long-term lease or build) globally ranging dedicated vessel.

For example, the current U.S.-dedicated drillship (*JOIDES Resolution*) model differs in important ways from a potential future U.S. MSP model. In the current dedicated drillship model, expedition scheduling employs a regional planning approach to save cost and time. Transparent regional planning by facility boards allows proponent teams to develop proposals in support of strategically timed scientific ocean drilling in a particular area of the global ocean. A disadvantage of this approach, however, may be that all expeditions are not contributing to the highest-priority science goals. Instead, they may be designed around what is geographically convenient, not what is most important. MSP operations have more geographic flexibility but are often complex to implement in terms of planning, since the planning is indeed mission specific and involves third parties, which increases complexity (and can decrease efficiency and increase risk and cost). While MSPs provide an opportunity to achieve some of the focused science objectives, fewer science objectives would likely be achieved.

At this critical juncture, it would be appropriate for NSF to undertake an assessment to determine essential aspects of the management and staffing structure, including the advisory structure, of shore-based and platform-based facilities. This section continues with a list of potential generative questions and subjects that could be considered in such an assessment. The committee suggests these questions, not to provide a specific prescription, but rather to signal the scale and scope of management infrastructure considerations.

- **Platform options:** What are the short- and long-term financial, scientific, operational, and leadership advantages and disadvantages of developing a U.S. MSP-type program versus acquiring a dedicated globally ranging drilling vessel? Are there existing appropriate drilling vessels available for long-term lease or is a new build necessary? What are the advantages and disadvantages of a full industry contract/lead option versus a full NSF or nonprofit partnership option? How could the ARF be enhanced to provide additional options, such as for seafloor coring (e.g., via giant piston coring) and drilling (e.g., via seabed lander drilling systems)?

- **Platform-based measurements:** Which measurements are required to be made on the platform (e.g., ephemeral, required for drilling decisions)? Fewer measurements taken means simpler operations, fewer staff, less travel, less shore-based administrative support, among other impacts. What is the absolute minimum needed for materials recovered? The answer to this fundamental question has a cascade of implications. Where then would these measurements be made? Would an existing U.S. repository be redesigned to provide such laboratory space (similar to how the laboratories associated with the Bremen Core Repository support MSP expeditions)?
- **Shore-based scientific and technical support:** With fewer measurements made at sea, what is the justification for shore-based, nonrepository staff? Is the current model of supporting expedition project managers (EPOs) (formerly called staff scientists) still relevant (vs. a more distributed model, for example)? Would technicians be needed to support and run a repository-based suite of laboratories; would EPOs be needed to manage LEAPs as well as future drilling expeditions? Which support operations can be contracted to a third party?
- **Advisory structure:** What is the minimum advisory structure needed? With so many highly ranked research proposals stored in the program, how many new ones are needed? Do the existing proposals already seek to address high-priority science identified in this report and in the *2050 Framework* that would fill in critical gaps (temporally and spatially)? Rather than a continually empaneled selection committee, should proposal selection (and other advisory functions) be moved to a biennial (or less frequent) basis? What aspects of the planning process can be moved to a multiyear basis?

The answers to these questions could help trim operations and maintenance costs, which have plagued the ocean sciences community for not only scientific drilling, but other critical infrastructure as well (see *Sea Change: 2015-2025 Decadal Survey of Ocean Sciences* [NRC, 2015]). **Identifying the minimum required program capabilities in order to advance vital and urgent scientific goals would help to facilitate a stable, successful, dynamic, and sustainable U.S. scientific ocean drilling research program.**

**CONCLUSION 4.3** Scientific ocean drilling is now at a critical juncture; the future of scientific ocean drilling itself and progress on globally urgent and vital research is at risk if U.S. operational leadership and participation end. Some high-priority science questions, with the potential to yield societal benefits, are best addressed and can only be addressed with ocean drilling research. Advancing that research requires consideration for new approaches to address resource, infrastructure, and capacity needs.

# References


Aagaard, B., M. Knepley, and C. WIlliams. 2022. *PyLith, v3.0.3*. Computational Infrastructure for Geodynamics. https://doi.org/10.5281/zenodo.7072811.

Andreani, M., G. Montagnac, C. Fellah, J. Hao, F. Vandier, I. Daniel, C. Pisapia, J. Galipaud, M. D. Lilley, G. L. Früh Green, S. Borensztajn, and B. Ménez. 2023. The rocky road to organics needs drying. *Nature Communications* 14(1):347. https://doi.org/10.1038/s41467-023-36038-6.

Araki, E., D. M. Saffer, A. J. Kopf, L. M. Wallace, T. Kimura, Y. Machida, S. Ide, E. Davis, and IODP Expedition 365 Shipboard Scientists. 2017. Recurring and triggered slow-slip events near the trench at the Nankai Trough subduction megathrust. *Science* 356(6343):1157-1160. https://doi.org/doi:10.1126/science.aan3120.

Arculus, R. J., O. Ishizuka, K. A. Bogus, M. Gurnis, R. Hickey-Vargas, M. H. Aljahdali, A. N. Bandini-Maeder, A. P. Barth, P. A. Brandl, L. Drab, R. do Monte Guerra, M. Hamada, F. Jiang, K. Kanayama, S. Kender, Y. Kusano, H. Li, L. C. Loudin, M. Maffione, K. M. Marsaglia, A. McCarthy, S. Meffre, A. Morris, M. Neuhaus, I. P. Savov, C. Sena, F. J. Tepley III, C. van der Land, G. M. Yogodzinski, and Z. Zhang. 2015. A record of spontaneous subduction initiation in the Izu–Bonin–Mariana arc. *Nature Geoscience* 8(9):728-733. https://doi.org/10.1038/ngeo2515.

Auderset, A., S. Moretti, B. Taphorn, P.-R. Ebner, E. Kast, X. T. Wang, R. Schiebel, D. M. Sigman, G. H. Haug, and A. Martínez-García. 2022. Enhanced ocean oxygenation during Cenozoic warm periods. *Nature* 609(7925):77-82. https://doi.org/10.1038/s41586-022-05017-0.

Baldauf, J., K. Becker, E. Davis, P. J. Fox, K. K. Graber, K. Grigar, R. Grout, B. Jonasson, E. Pollard, E. Schulte, and D. Schroeder. 2002. Advanced piston corer. In *Overview of ocean drilling program engineering tools and hardware*, edited by K. K. Graber, E. Pollard, B. Jonasson, and E. Schulte. College Station, TX: ODP Publications. http://www-odp.tamu.edu/publications/tnotes/tn31/apc/apc.htm.

Bar-On, Y. M., R. Phillips, and R. Milo. 2018. The biomass distribution on Earth. *Proceedings of the National Academy of Sciences* 115(25):6506-6511. https://doi.org/doi:10.1073/pnas.1711842115.

Bartlow, N., L. M. Wallace, J. Elliott, and S. Schwartz. 2021. Slipping and locking in Earth's earthquake factories. *Eos* 102. https://doi.org/ https://doi.org/10.1029/2021EO155885. https://eos.org/science-updates/slipping-and-locking-in-earths-earthquake-factories.

Beulig, F., F. Schubert, R. R. Adhikari, C. Glombitza, V. B. Heuer, K. U. Hinrichs, K. L. Homola, F. Inagaki, B. B. Jørgensen, J. Kallmeyer, S. J. E. Krause, Y. Morono, J. Sauvage, A. J. Spivack, and T. Treude. 2022. Rapid metabolism fosters microbial survival in the deep, hot subseafloor biosphere. *Nature Communications* 13(1):312. https://doi.org/10.1038/s41467-021-27802-7.

Bhadra, S. R., and R. Saraswat. 2022. A strong influence of the mid-Pleistocene transition on the monsoon and associated productivity in the Indian Ocean. *Quaternary Science Reviews* 295(1):e107761.

Böhm, E., J. Lippold, M. Gutjahr, M. Frank, P. Blaser, B. Antz, J. Fohlmeister, N. Frank, M. B. Andersen, and M. Deininger. 2015. Strong and deep Atlantic meridional overturning circulation during the last glacial cycle. *Nature* 517(7532):73-76. https://doi.org/10.1038/nature14059.

Boscolo-Galazzo, F., A. Jones, T. Dunkley Jones, K. A. Crichton, B. S. Wade, and P. N. Pearson. 2022. Late Neogene evolution of modern deep-dwelling plankton. *Biogeosciences* 19(3):743-762. https://doi.org/10.5194/bg-19-743-2022.

Brahim, Y. A., M. C. Peros, A. E. Viau, M. Liedtke, J. M. Pajón, J. Valdes, X. Li, R. L. Edwards, E. G. Reinhardt, and F. Oliva. 2022. Hydroclimate variability in the Caribbean during North Atlantic Heinrich cooling events (H8 and H9). *Scientific Reports* 12(1):e20719. https://doi.org/10.1038/s41598-022-24610-x.

Brinkhuis, H. 2023. Progress on the long-term objectives of the International Ocean Discovery Program Science Plan (and associated issues). Presentation at the International Ocean Discovery Program Forum. https://www.iodp.org/docs/meetings/1206-chair-presentation-on-science-plan-fulfillment-2023/file.

Cai, L., B. B. Jørgensen, C. A. Suttle, M. He, B. A. Cragg, N. Jiao, and R. Zhang. 2019. Active and diverse viruses persist in the deep sub-seafloor sediments over thousands of years. *ISME Journal* 13(7):1857-1864. https://doi.org/10.1038/s41396-019-0397-9.

Camoin, G., and N. Eguchi. 2022. Post-2024 ECORD and Japan: Joint program planning. Presentation at IODP Forum Meeting, Toward Post-2024 Scientific Ocean Drilling, September 15, 2022 Discussion, September 14–15, 2022. https://www.iodp.org/docs/iodp-future/1192-post-2024-discussion-220915/file.

Camoin, G., and N. Eguchi. 2023. Japan and Europe's next ocean-drilling research programme. *Nature* 616(7955):33. https://doi.org/10.1038/d41586-023-00893-6.

Carmichael, M. J., D. J. Lunt, M. Huber, M. Heinemann, J. Kiehl, A. LeGrande, C. A. Loptson, C. D. Roberts, N. Sagoo, C. Shields, P. J. Valdes, A. Winguth, C. Winguth, and R. D. Pancost. 2016. A model–model and data–model comparison for the early Eocene hydrological cycle. *Climate of the Past* 12(2):455-481. https://doi.org/10.5194/cp-12-455-2016.

Carmichael, M. J., G. N. Inglis, M. P. S. Badger, B. D. A. Naafs, L. Behrooz, S. Remmelzwaal, F. M. Monteiro, M. Rohrssen, A. Farnsworth, H. L. Buss, A. J. Dickson, P. J. Valdes, D. J. Lunt, and R. D. Pancost. 2017. Hydrological and associated biogeochemical consequences of rapid global warming during the Paleocene-Eocene thermal maximum. *Global and Planetary Change* 157:114-138. https://doi.org/10.1016/j.gloplacha.2017.07.014.

Chu, M., R. Bao, M. Strasser, K. Ikehara, J. Everest, L. Maeda, K. Hochmuth, L. Xu, A. McNichol, P. Bellanova, T. Rasbury, M. Kölling, N. Riedinger, J. Johnson, M. Luo, C. März, S. Straub, K. Jitsuno, M. Brunet, Z. Cai, A. Cattaneo, K. Hsiung, T. Ishizawa, T. Itaki, T. Kanamatsu, M. Keep, A. Kioka, C. McHugh, A. Micallef, D. Pandey, J. N. Proust, Y. Satoguchi, D. Sawyer, C. Seibert, M. Silver, J. Virtasalo, Y. Wang, T.-W. Wu, and S. Zellers. 2023. Earthquake-enhanced dissolved carbon cycles in ultra-deep ocean sediments. *Nature Communications* 14(1):5427. https://doi.org/10.1038/s41467-023-41116-w.

Connock, G. T., J. D. Owens, and X.-L. Liu. 2022. Biotic induction and microbial ecological dynamics of Oceanic Anoxic Event 2. *Communications Earth & Environment* 3(1):136. https://doi.org/10.1038/s43247-022-00466-x.

Cotterill, C., D. Haas, A. Lam, and K. St John. 2021a. *Scientific ocean drilling: Preparing the next generation workshop summary*. International Ocean Discovery Program. https://serc.carleton.edu/iodp/2021-nextgen/summary.html.

Cotterill, C., K. Homola, R. Norris, S. O'Connell, and M. Schulte. 2021b. *Scientific ocean drilling: Workshop summary*. International Ocean Discovery Program. https://serc.carleton.edu/iodp/july2021-policy/summary.html.

Cramer, B. S., J. R. Toggweiler, J. D. Wright, M. E. Katz, and K. G. Miller. 2009. Ocean overturning since the Late Cretaceous: Inferences from a new benthic foraminiferal isotope compilation. *Paleoceanography* 24:PA2416. https://doi.org/10.1029/2008PA001683.

Crichton, K. A., J. D. Wilson, A. Ridgwell, F. Boscolo-Galazzo, E. H. John, B. S. Wade, and P. N. Pearson. 2023. What the geological past can tell us about the future of the ocean's twilight zone. *Nature Communications* 14(1):2376. https://doi.org/10.1038/s41467-023-37781-6.

Curry, W. B., and D. Oppo. 2005. Glacial water mass geometry and the distribution of $\delta^{13}C$ of $\Sigma CO_2$ in the western Atlantic Ocean. *Paleoceanography* 20:PA1017. https://doi.org/10.1029/2004PA001021.

DeConto, R. M., and D. Pollard. 2016. Contribution of Antarctica to past and future sea-level rise. *Nature* 531(7596):591-597. https://doi.org/10.1038/nature17145.

DeConto, R. M., D. Pollard, R. B. Alley, I. Velicogna, E. Gasson, N. Gomez, S. Sadai, A. Condron, D. M. Gilford, E. L. Ashe, R. E. Kopp, D. Li, and A. Dutton. 2021. The Paris Climate Agreement and future sea-level rise from Antarctica. *Nature* 593(7857):83-89. https://doi.org/10.1038/s41586-021-03427-0.

D'Hondt, S., G. Wang, and A. J. Spivack. 2014. The underground economy: Energetic constraints of subseafloor life. In *Developments in marine geology*, Vol. 7, edited by R. Stein, D. K. Blackman, F. Inagaki, and H.-C. Larsen. Elsevier. Pp. 127-148. https://doi.org/10.1016/B978-0-444-62617-2.00005-0.

D'Hondt, S., F. Inagaki, B. N. Orcutt, and K.-U. Hinrichs. 2019a. IODP advances in the understanding of subseafloor life. *Oceanography* 32(1):198-207. https://www.jstor.org/stable/26604977.

D'Hondt, S., R. Pockalny, V. M. Fulfer, and A. J. Spivack. 2019b. Subseafloor life and its biogeochemical impacts. *Nature Communications* 10(1):3519. https://doi.org/10.1038/s41467-019-11450-z.

Dick, H. J. B., A. J. S. Kvassnes, P. T. Robinson, C. J. MacLeod, and H. Kinoshita. 2019. The Atlantis Bank Gabbro Massif, Southwest Indian Ridge. *Progress in Earth and Planetary Science* 6(1):64. https://doi.org/10.1186/s40645-019-0307-9.

Ditlevsen, P., and S. Ditlevsen. 2023. Warning of a forthcoming collapse of the Atlantic meridional overturning circulation. *Nature Communications* 14(1):4254. https://doi.org/10.1038/s41467-023-39810-w.

Druffel, E. R. M., P. M. Williams, J. E. Bauer, and J. R. Ertel. 1992. Cycling of dissolved and particulate organic matter in the open ocean. *Journal of Geophysical Research* 97(C10):15639-15659.

Dutton, A., A. E. Carlson, A. J. Long, G. A. Milne, P. U. Clark, R. DeConto, B. P. Horton, S. Rahmstorf, and M. E. Raymo. 2015. Sea-level rise due to polar ice-sheet mass loss during past warm periods. *Science* 349(6244):aaa4019. https://doi.org/doi:10.1126/science.aaa4019.

ECORD (European Consortium for Ocean Research Drilling). 2023. *The International Ocean Drilling Programme-3 (IODP$^3$) will start on January 1st, 2025*. https://www.ecord.org/the-international-ocean-drilling-programme-3-iodp-cubed-will-start-on-january-1st-2025/.

Edwards, K., W. Bach, A. Klaus, and the IODP Expedition 336 Scientific Party. 2014. IODP Expedition 336: Initiation of long-term coupled microbiological, geochemical, and hydrological experimentation within the seafloor at North Pond, western flank of the Mid-Atlantic Ridge. *Scientific Drilling* 17:13-18. https://doi.org/10.5194/sd-17-13-2014.

Ekpo Johnson, E., M. Scherwath, K. Moran, S. E. Dosso, and K. M. Rohr. 2023. Fault slip tendency analysis for a deep-sea basalt $CO_2$ injection in the Cascadia Basin. *GeoHazards* 4(2):121-135. https://doi.org/10.3390/geohazards4020008.

Executive Office of the President (EOP). 2023. Ocean Climate Action Plan: A Report by the Ocean Policy Committee.

Fisher, A. T. C. G. Wheat, K. Becker, J. Cowen, B. Orcutt, S. Hulme, K. Inderbitzen, A. Haddad, T. L. Pettigrew, E. E. Davis, H. Jannasch, K. Grigar, R. Aduddell, R. Meldrum, R. Macdonald, and K. J. Edwards. 2011. Design, deployment, and status of borehole observatory systems used for single-hole and cross-hole experiments, IODP Expedition 327, eastern flank of Juan de Fuca Ridge. Proceedings of the Integrated Ocean Drilling Program, Vol. 327. http://publications.iodp.org/proceedings/327/107/107_f2.htm.

Früh-Green, G. L., B. Orcutt, S. Green, C. Cotterill, and Expedition 357 Scientists. 2015. Atlantis Massif serpentinization and life. *Proceedings of the International Ocean Discovery Program* 357. http://publications.iodp.org/proceedings/357/357title.html.

Geilert, S., P. Grasse, K. Wallmann, V. Liebetrau, and C. D. Menzies. 2020. Serpentine alteration as source of high dissolved silicon and elevated $\delta^{30}$Si values to the marine Si cycle. *Nature Communications* 11(1):5123. https://doi.org/10.1038/s41467-020-18804-y.

GIDA (Global Indigenous Data Alliance). n.d. CAR*E principles for Indigenous data governance*. https://www.gida-global.org/care.

Gohl, K., G. Uenzelmann-Neben, J. Gille-Petzoldt, C.-D. Hillenbrand, J. P. Klages, S. M. Bohaty, S. Passchier, T. Frederichs, J. S. Wellner, R. Lamb, G. Leitchenkov, and IODP Expedition 379 Scientists. 2021. Evidence for a highly dynamic West Antarctic ice sheet during the Pliocene. *Geophysical Research Letters* 48(14):e2021GL093103. https://doi.org/https://doi.org/10.1029/2021GL093103.

Goldberg, D., and A. L. Slagle. 2009. A global assessment of deep-sea basalt sites for carbon sequestration. *Energy Procedia* 1(1):3675-3682.

Goldfinger, C., C. H. Nelson, A. E. Morey, J. E. Johnson, J. R. Patton, E. B. Karabanov, J. Gutierrez-Pastor, A. T. Eriksson, E. Gracia, G. Dunhill, R. J. Enkin, A. Dallimore, and T. Vallier. 2012. *Turbidite event history: Methods and implications for Holocene paleoseismicity of the Cascadia subduction zone*. Professional Paper 1661-F. U.S. Geological Survey. https://pubs.usgs.gov/publication/pp1661F.

Gulev, S. K., P. W. Thorne, J. Ahn, F. J. Dentener, C. M. Domingues, S. Gerland, D. Gong, D. S. Kaufman, H. C. Nnamchi, J. Quaas, J. A. Rivera, S. Sathyendranath, S. L. Smith, B. Trewin, K. von Schuckmann, and R. S. Vose. 2021. Changing state of the climate system. In *Climate Change 2021: The Physical Science Basis. Contribution of Working Group I to the Sixth Assessment Report of the Intergovernmental Panel on Climate Change*, edited by V. Masson-Delmotte, P. Zhai, A. Pirani, S. L. Connors, C. Péan, S. Berger, N. Caud, Y. Chen, L. Goldfarb, M. I. Gomis, M. Huang, K. Leitzell, E. Lonnoy, J. B. R. Matthews, T. K. Maycock, T. Waterfield, O. Yelekçi, R. Yu, and B. Zhou. Cambridge: Cambridge University Press. Pp. 287-422. https://doi.com/10.1017/9781009157896.004.

Halpern, B. S., C. Longo, J. S. S. Lowndes, B. D. Best, M. Frazier, S. K. Katona, K. M. Kleisner, A. A. Rosenberg, C. Scarborough, and E. R. Selig. 2015. Patterns and emerging trends in global ocean health. *PLoS ONE* 10(3):e0117863. https://doi.org/10.1371/journal.pone.0117863.

Hedges, J. I. 1992. Global biogeochemical cycles: Progress and problems. *Marine Chemistry* 39(1-3):67-93.

Held, I. M., and B. J. Soden. 2006. Robust responses of the hydrological cycle to global warming. *Journal of Climate* 19(21):5686-5699. https://doi.org/https://doi.org/10.1175/JCLI3990.1.

Henson, S. A., C. Laufkötter, S. Leung, S. L. C. Giering, H. I. Palevsky, and E. L. Cavan. 2022. Uncertain response of ocean biological carbon export in a changing world. *Nature Geoscience* 15(4):248-254. https://doi.org/10.1038/s41561-022-00927-0.

Heuer, V. B., F. Inagaki, Y. Morono, Y. Kubo, A. J. Spivack, B. Viehweger, T. Treude, F. Beulig, F. Schubotz, S. Tonai, S. A. Bowden, M. Cramm, S. Henkel, T. Hirose, K. Homola, T. Hoshino, A. Ijiri, H. Imachi, N. Kamiya, M. Kaneko, L. Lagostina, H. Manners, H.-L. McClelland, K. Metcalfe, N. Okutsu, D. Pan, M. J. Raudsepp, J. Sauvage, M.-Y. Tsang, D. T. Wang, E. Whitaker, Y. Yamamoto, K. Yang, L. Maeda, R. R. Adhikari, C. Glombitza, Y. Hamada, J. Kallmeyer, J. Wendt, L. Wörmer, Y. Yamada, M. Kinoshita, and K.-U. Hinrichs. 2020. Temperature limits to deep subseafloor life in the Nankai Trough subduction zone. *Science* 370(6521):1230-1234. https://doi.org/doi:10.1126/science.abd7934.

Hoehler, T. M., and B. B. Jørgensen. 2013. Microbial life under extreme energy limitation. *Nature Reviews Microbiology* 11(2):83-94. https://doi.org/10.1038/nrmicro2939.

Hollis, C. J., T. Dunkley Jones, E. Anagnostou, P. K. Bijl, M. J. Cramwinckel, Y. Cui, G. R. Dickens, K. M. Edgar, Y. Eley, D. Evans, G. L. Foster, J. Frieling, G. N. Inglis, E. M. Kennedy, R. Kozdon, V. Lauretano, C. H. Lear, K. Littler, L. Lourens, A. N. Meckler, B. D. A. Naafs, H. Pälike, R. D. Pancost, P. N. Pearson, U. Röhl, D. L. Royer, U. Salzmann, B. A. Schubert, H. Seebeck, A. Sluijs, R. P. Speijer, P. Stassen, J. Tierney, A. Tripati, B. Wade, T. Westerhold, C. Witkowski, J. C. Zachos, Y. G. Zhang, M. Huber, and D. J. Lunt. 2019. The DeepMIP contribution to PMIP4: Methodologies for selection, compilation and analysis of latest Paleocene and early Eocene climate proxy data, incorporating version 0.1 of the DeepMIP database. *Geoscientific Model Development* 12(7):3149-3206. https://doi.org/10.5194/gmd-12-3149-2019.

Hutchins, D. A., and D. G. Capone. 2022. The marine nitrogen cycle: New developments and global change. *Nature Reviews Microbiology* 20(7):401-414. https://doi.org/10.1038/s41579-022-00687-z.

IODP (International Ocean Discovery Program). n.d.a. *Coring tools and technology*. https://iodp.tamu.edu/tools/index.html.

IODP. n.d.b. *Countries participating in IODP.* https://www.iodp.org/about-iodp/countries-participating-in-iodp.

IODP. n.d.c. *Exploring the Earth under the sea: Expedition schedule*. n.d.c. https://www.iodp.org/expeditions/expeditions-schedule.

IODP. n.d.d. *Exploring the Earth under the sea: Expedition statistics*. https://www.iodp.org/expeditions/expedition-statistics.

IODP. n.d.e. *Exploring the Earth under the sea: IODP core repositories*. https://www.iodp.org/resources/core-repositories.

IODP. n.d.f. *Facility assessments*. https://iodp.tamu.edu/publications/assessment.html.

IODP. n.d.g. *HRT animation*. https://iodp.tamu.edu/tools/video/IODP_HRT_Tool_Animation.mp4.

IODP. n.d.h. JOIDES Resolution *School of Rock*. https://joidesresolution.org/for-educators/school-of-rock.

IODP. n.d.i. *Laboratories*. https://iodp.tamu.edu/labs/index.html.

IODP. n.d.j. *Planning workshop outcomes*. https://www.iodp.org/iodp-future/planning-workshop-outcomes.

IODP. n.d.k. *Technology and tools*. https://iodp.merlinone.net/MX/ContentHub/Digital_Library/index.html?portalview=8613&assetview=161049.

IODP. n.d.l. *Tool animation*. https://iodp.tamu.edu/tools/video/IODP_APC_Tool_Animation.m4v.

IODP. 2001. *Earth, oceans, and life: Scientific investigation of the Earth system using multiple drilling platforms and new technologies: Integrated Ocean Drilling Program initial science plan, 2003–2013*. https://www.iodp.org/iodp-legacy/iodp-2003-2013-documents/iodp-2003-2013-general-reference-documents/735-iodp-2003-2013-initial-science-plan/file.

IODP. 2011. *2013–2023 IODP Science Plan: Illuminating Earth's past, present, and future*. https://www.iodp.org/about-iodp/iodp-science-plan-2013–2023.

IODP. 2019. *Proceedings of the International Ocean Discovery Program 385: Guaymas Basin tectonics and biosphere*. http://publications.iodp.org/proceedings/385/385title.html.

Jørgensen, B. B., and I. P. G. Marshall. 2016. Slow microbial life in the seabed. *Annual Review of Marine Science* 8(1):311-332. https://doi.org/10.1146/annurev-marine-010814-015535.

Jørgensen, B. B., T. Andrén, and I. P. G. Marshall. 2020. Sub-seafloor biogeochemical processes and microbial life in the Baltic Sea. *Environmental Microbiology* 22(5):1688-1706. https://doi.org/10.1111/1462-2920.14920.

Kallmeyer, J., R. Pockalny, R. R. Adhikari, D. C. Smith, and S. D'Hondt. 2012. Global distribution of microbial abundance and biomass in subseafloor sediment. *Proceedings of the National Academy of Sciences* 109(40):16213-16216. https://doi.org/doi:10.1073/pnas.1203849109.

Kappel, E., ed. 2019. Special issue on scientific ocean drilling: Looking to the future in oceanography. *Oceanography* 32(1). https://tos.org/oceanography/issue/volume-32-issue-01.

Kast, E. R., D. A. Stolper, A. Auderset, J. A. Higgins, H. Ren, X. T. Wang, A. Martínez-García, G. H. Haug, and D. M. Sigman. 2019. Nitrogen isotope evidence for expanded ocean suboxia in the early Cenozoic. *Science* 364(6438):386-389. https://doi.org/doi:10.1126/science.aau5784.

Kennett, J. P., A. R. McBirney, and R. C. Thunell. 1977. Episodes of Cenozoic volcanism in the circum-Pacific region. *Journal of Volcanology and Geothermal Research* 2(2):145-163. https://doi.org/https://doi.org/10.1016/0377-0273(77)90007-5.

Kikuchi, K., K. Abiko, H. Nagahama, and J. Muto. 2014. Self-affinities analysis of fault-related folding. *International Union of Geological Sciences* 37(4):308-311. https://doi.org/10.18814/epiiugs/2014/v37i4/011.

Kinkel, H. 2023, May 19. *The International Ocean Drilling Programme-3 (IODP-cubed) will start on January 1st, 2025.* European Consortium for Ocean Drilling Research. https://www.ecord.org/the-international-ocean-drilling-programme-3-iodp-cubed-will-start-on-january-1st-2025/

Kopf, A., T. Freudenthal, V. Ratmeyer, M. Bergenthal, M. Lange, T. Fleischmann, S. Hammerschmidt, C. Seiter, and G. Wefer. 2015. Simple, affordable, and sustainable borehole observatories for complex monitoring objectives. *Geoscientific Instrumentation, Methods, and Data Systems* 4(1):99-109. https://doi.org/10.5194/gi-4-99-2015.

Koppers, A. A. P., and R. Coggon, eds. 2020. *2050 Science framework: Exploring Earth by scientific ocean drilling*. International Ocean Discovery Program. https://www.iodp.org/2050-science-framework.

Lam, A. R., and R. M. Leckie. 2020. Late Neogene and Quaternary diversity and taxonomy of subtropical to temperate planktic foraminifera across the Kuroshio Current Extension, northwest Pacific Ocean. *Micropaleontology*, 66(3):177-268.

Larsen, H. C., G. Mohn, M. Nirrengarten, Z. Sun, J. Stock, Z. Jian, A. Klaus, C. A. Alvarez-Zarikian, J. Boaga, S. A. Bowden, A. Briais, Y. Chen, D. Cukur, K. Dadd, W. Ding, M. Dorais, E. C. Ferré, F. Ferreira, A. Furusawa, A. Gewecke, J. Hinojosa, T. W. Höfig, K. H. Hsiung, B. Huang, E. Huang, X. L. Huang, S. Jiang, H. Jin, B. G. Johnson, R. M. Kurzawski, C. Lei, B. Li, L. Li, Y. Li, J. Lin, C. Liu, C. Liu, Z. Liu, A. J. Luna, C. Lupi, A. McCarthy, L. Ningthoujam, N. Osono, D. W. Peate, P. Persaud, N. Qiu, C. Robinson, S. Satolli, I. Sauermilch, J. C. Schindlbeck, S. Skinner, S. Straub, X. Su, C. Su, L. Tian, F. M. van der Zwan, S. Wan, H. Wu, R. Xiang, R. Yadav, L. Yi, P. S. Yu, C. Zhang, J. Zhang, Y. Zhang, N. Zhao, G. Zhong, and L. Zhong. 2018. Rapid transition from continental breakup to igneous oceanic crust in the South China Sea. *Nature Geoscience* 11(10):782-789. https://doi.org/10.1038/s41561-018-0198-1.

Lever, M. A., O. Rouxel, J. C. Alt, N. Shimizu, S. Ono, R. M. Coggon, W. C. Shanks, L. Lapham, M. Elvert, X. Prieto-Mollar, K.-U. Hinrichs, F. Inagaki, and A. Teske. 2013. Evidence for microbial carbon and sulfur cycling in deeply buried ridge flank basalt. *Science* 339(6125):1305-1308. https://doi.org/10.1126/science.1229240.

Li, F., M. S. Lozier, S. Bacon, A. S. Bower, S. A. Cunningham, M. F. de Jong, B. deYoung, N. Fraser, N. Fried, G. Han, N. P. Holliday, J. Holte, L. Houpert, M. E. Inall, W. E. Johns, S. Jones, C. Johnson, J. Karstensen, I. A. Le Bras, P. Lherminier, X. Lin, H. Mercier, M. Oltmanns, A. Pacini, T. Petit, R. S. Pickart, D. Rayner, F. Straneo, V. Thierry, M. Visbeck, I. Yashayaev, and C. Zhou. 2021. Subpolar North Atlantic western boundary density anomalies and the Meridional Overturning Circulation. *Nature Communications* 12(1):3002. https://doi.org/10.1038/s41467-021-23350-2.

Li, H.-Y., R.-P. Zhao, J. Li, Y. Tamura, C. Spencer, R. J. Stern, J. G. Ryan, and Y.-G. Xu. 2021. Molybdenum isotopes unmask slab dehydration and melting beneath the Mariana arc. *Nature Communications* 12(1):6015. https://doi.org/10.1038/s41467-021-26322-8.

Li, L., S. Bai, J. Li, S. Wang, L. Tang, S. Dasgupta, Y. Tang, and X. Peng. 2020. Volcanic ash inputs enhance the deep-sea seabed metal-biogeochemical cycle: A case study in the Yap Trench, western Pacific Ocean. *Marine Geology* 430:e106340.

Lomstein, B. A., A. T. Langerhuus, S. D'Hondt, B. B. Jørgensen, and A. J. Spivack. 2012. Endospore abundance, microbial growth and necromass turnover in deep sub-seafloor sediment. *Nature* 484(7392):101-104. https://doi.org/10.1038/nature10905.

Lowery, C. M., T. J. Bralower, J. D. Owens, F. J. Rodríguez-Tovar, H. Jones, J. Smit, M. T. Whalen, P. Claeys, K. Farley, S. P. S. Gulick, J. V. Morgan, S. Green, E. Chenot, G. L. Christeson, C. S. Cockell, M. J. L. Coolen, L. Ferrière, C. Gebhardt, K. Goto, D. A. Kring, J. Lofi, R. Ocampo-Torres, L. Perez-Cruz, A. E. Pickersgill, M. H. Poelchau, A. S. P. Rae, C. Rasmussen, M. Rebolledo-Vieyra, U. Riller, H. Sato, S. M. Tikoo, N. Tomioka, J. Urrutia-Fucugauchi, J. Vellekoop, A. Wittmann, L. Xiao, K. E. Yamaguchi, and W. Zylberman. 2018. Rapid recovery of life at ground zero of the end-Cretaceous mass extinction. *Nature* 558(7709):288-291. https://doi.org/10.1038/s41586-018-0163-6.

Lunt, D. J., M. Huber, E. Anagnostou, M. L. J. Baatsen, R. Caballero, R. DeConto, H. A. Dijkstra, Y. Donnadieu, D. Evans, R. Feng, G. L. Foster, E. Gasson, A. S. von der Heydt, C. J. Hollis, G. N. Inglis, S. M. Jones, J. Kiehl, S. Kirtland Turner, R. L. Korty, R. Kozdon, S. Krishnan, J. B. Ladant, P. Langebroek, C. H. Lear, A. N. LeGrande, K. Littler, P. Markwick, B. Otto-Bliesner, P. Pearson, C. J. Poulsen, U. Salzmann, C. Shields, K. Snell, M. Stärz, J. Super, C. Tabor, J. E. Tierney, G. J. L. Tourte, A. Tripati, G. R. Upchurch, B. S. Wade, S. L. Wing, A. M. E. Winguth, N. M. Wright, J. C. Zachos, and R. E. Zeebe. 2017. The DeepMIP contribution to PMIP4: Experimental design for model simulations of the EECO, PETM, and pre-PETM (version 1.0). *Geoscientific Model Development* 10(2):889-901. https://doi.org/10.5194/gmd-10-889-2017.

Marschalek, J. W., L. Zurli, F. Talarico, T. van de Flierdt, P. Vermeesch, A. Carter, F. Beny, V. Bout-Roumazeilles, F. Sangiorgi, S. R. Hemming, L. F. Pérez, F. Colleoni, J. G. Prebble, T. E. van Peer, M. Perotti, A. E. Shevenell, I. Browne, D. K. Kulhanek, R. Levy, D. Harwood, N. B. Sullivan, S. R. Meyers, E. M. Griffith, C. D. Hillenbrand, E. Gasson, M. J. Siegert, B. Keisling, K. J. Licht, G. Kuhn, J. P. Dodd, C. Boshuis, L. De Santis, R. M. McKay, and IODP Expedition 374. 2021. A large West Antarctic Ice Sheet explains early Neogene sea-level amplitude. *Nature* 600(7889):450-455. https://doi.org/10.1038/s41586-021-04148-0.

MARUM Center for Marine Environmental Sciences. n.d.a. *IODP Bremen core repository*. https://www.marum.de/en/Research/IODP-Bremen-Core-Repository.html.

MARUM. n.d.b. *Sea-floor drill rig MARUM-MeBo20*. https://www.marum.de/en/Infrastructure/Sea-floor-drill-rig-MARUM-MeBo200.html.

McKay, D. I. A., A. Staal, J. F. Abrams, R. Winkelmann, B. Sakschewski, S. Loriani, I. Fetzer, S. E. Cornell, J. Rockström, and T. M. Lenton. 2022. Exceeding 1.5°C global warming could trigger multiple climate tipping points. *Science* 377(6611):eabn7950. https://doi.org/doi:10.1126/science.abn7950.

McKay, R. M., L. De Santis, and D. K. Kulhanek. 2019. Ross Sea West Antarctic ice sheet history. *Proceedings of the International Ocean Discovery Program* 374. College Station, TX: International Ocean Discovery Program. http://publications.iodp.org/proceedings/374/374title.html

Meinshausen, M., Z. R. J. Nicholls, J. Lewis, M. J. Gidden, E. Vogel, M. Freund, U. Beyerle, C. Gessner, A. Nauels, N. Bauer, J. G. Canadell, J. S. Daniel, A. John, P. B. Krummel, G. Luderer, N. Meinshausen, S. A. Montzka, P. J. Rayner, S. Reimann, S. J. Smith, M. van den Berg, G. J. M. Velders, M. K. Vollmer, and R. H. J. Wang. 2020. The shared socio-economic pathway (SSP) greenhouse gas concentrations and their extensions to 2500. *Geoscientific Model Development* 13(8):3571-3605. https://doi.org/10.5194/gmd-13-3571-2020.

Morgan, J., S. Gulick, C. L. Mellett, S. L. Green, and Expedition 364 Scientists. 2016. Chicxulub: Drilling the K-Pg impact crater. *Proceedings of the International Ocean Discovery Program* 364. College Station, TX: International Ocean Discovery Program. http://publications.iodp.org/proceedings/364/364title.html.

NASEM (National Academies of Sciences, Engineering, and Medicine). 2022a. *Cross-cutting themes for U.S. contributions to the UN Ocean Decade*. Washington, DC: The National Academies Press.

NASEM. 2022b. *Next generation Earth systems science at the National Science Foundation*. Washington, DC: The National Academies Press.

NASEM. 2022c. *A research strategy for ocean-based carbon dioxide removal and sequestration*. Washington, DC: The National Academies Press.

Norris, R. D., D. Kroon, and A. Klaus. 1998. Leg 171B. *Proceedings of the Ocean Drilling Program, Initial Reports* 171B. College Station, TX: Ocean Drilling Program. http://www-odp.tamu.edu/publications/citations/cite171b.html.

Norris, R. D., S. K. Turner, P. M. Hull, and A. Ridgwell. 2013. Marine ecosystem responses to Cenozoic global change. *Science* 341(6145):492-498. https://doi.org/doi:10.1126/science.1240543.

NRC (National Research Council). 2015. *Sea change: 2015-2025 Decadal survey of ocean sciences*. Washington, DC: The National Academies Press.

NSF (National Science Foundation). 2023. *Announcement of the non-renewal of the JOIDES Resolution Operations and Maintenance Cooperative Agreement*. https://www.nsf.gov/news/news_summ.jsp?cntn_id=306986&org=OCE (accessed September 26, 2023).

Orcutt, B., D. LaRowe, J. Biddle, F. Colwell, B. Glazer, B. Reese, J. Kirkpatrick, L. Lapham, H. Mills, J. Sylvan, S. Wankel, and C. Wheat. 2013. Microbial activity in the marine deep biosphere: Progress and prospects. *Frontiers in Microbiology* 4. https://doi.org/10.3389/fmicb.2013.00189.

Parkes, R. J., B. A. Cragg, S. J. Bale, J. M. Getliff, K. Goodman, P. A. Rochelle, J. C. Fry, A. J. Weightman, and S. M. Harvey. 1994. Deep bacterial biosphere in Pacific Ocean sediments. *Nature* 371(6496):410-413. https://doi.org/10.1038/371410a0.

Petrizzo, M. R., K. G. MacLeod, D. K. Watkins, E. Wolfgring, and B. T. Huber. 2022. Late Cretaceous paleoceanographic evolution and the onset of cooling in the Santonian at southern high latitudes (IODP Site U1513, SE Indian Ocean). *Paleoceanography and Paleoclimatology*, 37(1):e2021PA004353.

Rae, J. W. B., Y. G. Zhang, X. Liu, G. L. Foster, H. M. Stoll, and R. D. M. Whiteford. 2021. Atmospheric $CO_2$ over the past 66 million years from marine archives. *Annual Review of Earth and Planetary Sciences* 49(1):609-641. https://doi.org/10.1146/annurev-earth-082420-063026.

Rahmstorf, S. 2002. Ocean circulation and climate during the past 120,000 years. *Nature* 419:207-214.

Ranghieri, F., and M. Ishiwatari. 2014. *Learning from megadisasters: Lessons from the Great East Japan earthquake*. Washington, DC: World Bank. http://hdl.handle.net/10986/18864.

Reese, B. K., L. A. Zinke, M. S. Sobol, D. E. LaRowe, B. N. Orcutt, X. Zhang, U. Jaekel, F. Wang, T. Dittmar, D. Defforey, and B. Tully. 2018. Nitrogen cycling of active bacteria within oligotrophic sediment of the Mid-Atlantic Ridge flank. *Geomicrobiology Journal* 35(6):468-483.

Reimann, L., A. T. Vafeidis, and L. E. Honsel. 2023. Population development as a driver of coastal risk: Current trends and future pathways. *Cambridge Prisms: Coastal Futures* 1:e14. https://doi.org/10.1017/cft.2023.3.

Reysenbach, A.-L., E. St. John, J. Meneghin, G. E. Flores, M. Podar, N. Dombrowski, A. Spang, S. L'Haridon, S. E. Humphris, C. E. J. de Ronde, F. Caratori Tontini, M. Tivey, V. K. Stucker, L. C. Stewart, A. Diehl, and W. Bach. 2020. Complex subsurface hydrothermal fluid mixing at a submarine arc volcano supports distinct and highly diverse microbial communities. *Proceedings of the National Academy of Sciences* 117(51):32627-32638. https://doi.org/doi:10.1073/pnas.2019021117.

Rice, J. C., and M.-J. Rochet. 2005. A framework for selecting a suite of indicators for fisheries management. *ICES Journal of Marine Science* 62(3):516-527. https://doi.org/10.1016/j.icesjms.2005.01.003.

Roberts, N., A. Piotrowski, J. McManus, and L. Keigwin. 2010. Synchronous deglacial overturning and water mass source changes. *Science* 327:75-78.

Robinson, R. S., B. Dugan, C. Brenner, L. Krissek, S. Carr, T. S. Collett, J. P. Dodd, P. Fryer, P. M. Fulton, S. P. S. Gulick, H. Kitajima, A. A. P. Koppers, B. Marcks, J. D. Miller, Y. Rosenthal, A. Slagle, M. Tominaga, M. E. Torres, and J. S. Wellner. 2022. *Science mission requirements for a globally ranging, riserless drilling vessel for US Scientific Ocean Drilling*. U.S. Science Support Program. https://usoceandiscovery.org/wp-content/uploads/2022/10/SMR-FINAL-report-9-29-22.pdf.

Rohling, E. J., A. Sluijs, H. A. Dijkstra, P. Köhler, R. S. W. van de Wal, A. S. von der Heydt, D. J. Beerling, A. Berger, P. K. Bijl, M. Crucifix, R. DeConto, S. S. Drijfhout, A. Fedorov, G. L. Foster, A. Ganopolski, J. Hansen, B. Hönisch, H. Hooghiemstra, M. Huber, P. Huybers, R. Knutti, D. W. Lea, L. J. Lourens, D. Lunt, V. Masson-Delmotte, M. Medina-Elizalde, B. Otto-Bliesner, M. Pagani, H. Pälike, H. Renssen, D. L. Royer, M. Siddall, P. Valdes, J. C. Zachos, R. E. Zeebe, and P. P. Members. 2012. Making sense of palaeoclimate sensitivity. *Nature* 491(7426):683-691. https://doi.org/10.1038/nature11574.

Rohling, E. J., G. Marino, G. L. Foster, P. A. Goodwin, A. S. von der Heydt, and P. Köhler. 2018. Comparing climate sensitivity, past and present. *Annual Review of Marine Science* 10(1):261-288. https://doi.org/10.1146/annurev-marine-121916-063242.

Sessa, J. A., A. J. Fraass, L. J. LeVay, K. M. Jamson, and S. E. Peters. 2023. The Extending Ocean Drilling Pursuits (eODP) Project: Synthesizing scientific ocean drilling data. *Geochemistry, Geophysics, Geosystems* 24(3):e2022GC010655. https://doi.org/10.1029/2022GC010655.

Shah Walter, S. R., U. Jaekel, H. Osterholz, A. T. Fisher, J. A. Huber, A. Pearson, T. Dittmar, and P. R. Girguis. 2018. Microbial decomposition of marine dissolved organic matter in cool oceanic crust. *Nature Geoscience* 11(5):334-339. https://doi.org/10.1038/s41561-018-0109-5.

Sherwood, S. C., M. J. Webb, J. D. Annan, K. C. Armour, P. M. Forster, J. C. Hargreaves, G. Hegerl, S. A. Klein, K. D. Marvel, E. J. Rohling, M. Watanabe, T. Andrews, P. Braconnot, C. S. Bretherton, G. L. Foster, Z. Hausfather, A. S. von der Heydt, R. Knutti, T. Mauritsen, J. R. Norris, C. Proistosescu, M. Rugenstein, G. A. Schmidt, K. B. Tokarska, and M. D. Zelinka. 2020. An assessment of Earth's climate sensitivity using multiple lines of evidence. *Reviews of Geophysics* 58(4):e2019RG000678. https://doi.org/https://doi.org/10.1029/2019RG000678.

Sigman, D. M., and E. A. Boyle. 2000. Glacial/interglacial variations in atmospheric carbon dioxide. *Nature* 407(6806):859-869. https://doi.org/10.1038/35038000.

Strasser, M., K. Ikehara, and C. Cotterill. 2019. *International Ocean Discovery Program Expedition 386 scientific prospectus: Japan Trench paleoseismology*. International Ocean Discovery Program. http://publications.iodp.org/scientific_prospectus/386/.

Taira, A., S. Toczko, N. Eguchi, S. Kuramoto, Y. Kubo, and W. Azuma. 2014. Recent scientific and operational achievements of D/V *Chikyu*. *Geoscience Letters* 1(1):2. https://doi.org/10.1186/2196-4092-1-2.

Thomalla, S. J., S.-A. Nicholson, T. J. Ryan-Keogh, and M. E. Smith. 2023. Widespread changes in Southern Ocean phytoplankton blooms linked to climate drivers. *Nature Climate Change* 13(9):975-984. https://doi.org/10.1038/s41558-023-01768-4.

Tittensor, D. P., C. Mora, W. Jetz, H. K. Lotze, D. Ricard, E. V. Berghe, and B. Worm. 2010. Global patterns and predictors of marine biodiversity across taxa. *Nature* 466(7310):1098-1101.

Tobin, H., T. Hirose, D. Saffer, S. Toczko, L. Maeda, Y. Kubo, B. Boston, A. Broderick, K. Brown, A. Crespo-Blanc, E. Even, S. Fuchida, R. Fukuchi, S. Hammerschmidt, P. Henry, M. Josh, M. J. Jurado, H. Kitajima, M. Kitamura, A. Maia, M. Otsubo, J. Sample, A. Schleicher, H. Sone, C. Song, R. Valdez, Y. Yamamoto, K. Yang, Y. Sanada, Y. Kido, and Y. Hamada. 2015. Expedition 348 summary. *Proceedings of the Integrated Ocean Drilling Program* 348. College Station, TX: Integrated Ocean Drilling Program.

Tobin, H. J., G. Kimura, and S. Kodaira. 2019. Processes governing giant subduction earthquakes: IODP drilling to sample and instrument subduction zone megathrusts. *Oceanography* 32(1):80-93.

Tobin, H., T. Hirose, M. Ikari, K. Kanagawa, G. Kimura, M. Kinoshita, H. Kitajima, D. Saffer, A. Yamaguchi, N. Eguchi, L. Maeda, S. Toczko, and the Expedition 358 Scientists. 2020. Expedition 358 Summary. *Proceedings of the International Ocean Discovery Program* 358. College Station, TX: International Ocean Discovery Program. https://doi.org/10.14379/iodp.proc.358.101.2020.

Toggweiler, J. R. 1999. Variation of atmospheric $CO_2$ by ventilation of the ocean's deepest water. *Paleoceanography* 14(5):571-588. https://doi.org/https://doi.org/10.1029/1999PA900033.

Trembath-Reichert, E., S. R. Shah Walter, M. A. F. Ortiz, P. D. Carter, P. R. Girguis, and J. A. Huber. 2021. Multiple carbon incorporation strategies support microbial survival in cold subseafloor crustal fluids. *Science Advances* 7(18):eabg0153. https://doi.org/doi:10.1126/sciadv.abg0153.

Trenberth, K. E. 2011. Changes in precipitation with climate change. *Climate Research* 47(1-2):123-138. https://www.int-res.com/abstracts/cr/v47/n1-2/p123-138.

Trenberth, K. E., and J. M. Caron. 2001. Estimates of meridional atmosphere and ocean heat transports. *Journal of Climate* 14(16):3433-3443. https://doi.org/10.1175/1520-0442(2001)014<3433:EOMAAO>2.0.CO;2.

Ujiie, K., and G. Kimura. 2014. Earthquake faulting in subduction zones: Insights from fault rocks in accretionary prisms. *Progress in Earth and Planetary Science* 1(1):7. https://doi.org/10.1186/2197-4284-1-7.

Varadi, F., B. Runnegar, and M. Ghil. 2003. Successive refinements in long-term integrations of planetary orbit. *Astrophysical Journal* 592:620-630. https://iopscience.iop.org/article/10.1086/375560/pdf.

Wade, B. S., P. N. Pearson, W. A. Berggen, and H. Pälike. 2011. Review and revision of Cenozoic tropical planktonic foraminiferal biostratigraphy and calibration to the geomagnetic polarity and astronomical time scale. *Earth-Science Reviews* 104:111-142. https://doi.org/10.1016/j.earscirev.2010.09.003.

Weber, M. E., I. Bailey, S. R. Hemming, Y. M. Martos, B. T. Reilly, T. A. Ronge, S. Brachfeld, T. Williams, M. Raymo, S. T. Belt, L. Smik, H. Vogel, V. L. Peck, L. Armbrecht, A. Cage, F. G. Cardillo, Z. Du, G. Fauth, C. J. Fogwill, M. Garcia, M. Garnsworthy, A. Glüder, M. Guitard, M. Gutjahr, I. Hernández-Almeida, F. S. Hoem, J.-H. Hwang, M. Iizuka, Y. Kato, B. Kenlee, S. Oconnell, L. F. Pérez, O. Seki, L. Stevens, L. Tauxe, S. Tripathi, J. Warnock, and X. Zheng. 2022. Antiphased dust deposition and productivity in the Antarctic Zone over 1.5 million years. *Nature Communications* 13(1):2044. https://doi.org/10.1038/s41467-022-29642-5.

Weijer, W., W. Cheng, S. S. Drijfhout, A. V. Fedorov, A. Hu, L. C. Jackson, W. Liu, E. L. McDonagh, J. V. Mecking, and J. Zhang. 2019. Stability of the Atlantic meridional overturning circulation: A review and synthesis. *Journal of Geophysical Research: Oceans* 124(8):5336-5375. https://doi.org/https://doi.org/10.1029/2019JC015083. https://agupubs.onlinelibrary.wiley.com/doi/abs/10.1029/2019JC015083.

White, L., S. Cooper, R. Wold-Brennon, and J. Lewis. 2021. *Engaging the public workshop summary.* Scientific Ocean Drilling. https://serc.carleton.edu/iodp/june2021-public/summary.html.

Winguth, A. M. E., E. Thomas, and C. Winguth. 2012. Global decline in ocean ventilation, oxygenation, and productivity during the Paleocene-Eocene thermal maximum: Implications for the benthic extinction. *Geology* 40(3):263-266. https://doi.org/10.1130/G32529.1.

Woodhouse, A., A. Swain, W. F. Fagan, A. J. Fraass, and C. M. Lowery. 2023. Late Cenozoic cooling restructured global marine plankton communities. *Nature* 614(7949):713-718. https://doi.org/10.1038/s41586-023-05694-5.

Yand, L., and L. Shumei. 2024. China's first homegrown ocean drillship completes trial voyage, set to make greater contributions to international scientific ocean drilling. *Global Times*. https://www.globaltimes.cn/page/202401/1305206.shtm.

Yasuhara, M., D. P. Tittensor, H. Hillebrand, and B. Worm. 2017. Combining marine macroecology and palaeoecology in understanding biodiversity: Microfossils as a model. *Biological Reviews* 92(1):199-215. https://doi.org/10.1111/brv.12223.

Zhang, Y., T. Huck, C. Lique, Y. Donnadieu, J. B. Ladant, M. Rabineau, and D. Aslanian. 2020. Early Eocene vigorous ocean overturning and its contribution to a warm Southern Ocean. *Climate of the Past* 16(4):1263-1283. https://doi.org/10.5194/cp-16-1263-2020.

Zhu, J., B. L. Otto-Bliesner, E. C. Brady, C. J. Poulsen, J. E. Tierney, M. Lofverstrom, and P. DiNezio. 2021. Assessment of equilibrium climate sensitivity of the Community Earth System Model Version 2 through simulation of the last glacial maximum. *Geophysical Research Letters* 48(3):e2020GL091220. https://doi.org/https://doi.org/10.1029/2020GL091220.

# Appendix A

# Committee Statement of Task

The Decadal Survey will advise the National Science Foundation's Division of Ocean Sciences (NSF OCE) on forward-looking approaches to guide investments in research, infrastructure, and workforce development. The committee will develop a compelling research and infrastructure strategy to advance understanding of the ocean's role in the Earth system and the sustainable blue economy. The report will recommend ways that NSF OCE could develop the capacity to respond nimbly as priorities change and new opportunities emerge over the 2025-2035 decade.

The committee will produce an interim report to provide advice to NSF OCE on the resources and infrastructure available to address high-priority research questions requiring scientific ocean drilling. The interim report will cover the following:

1. Based on previous reports, assess progress on addressing high-priority science questions that require scientific ocean drilling and identify new, if any, equally compelling science questions that would also require scientific ocean drilling.
2. Of the unanswered scientific questions, which could be addressed through the use of existing scientific drilling assets including sediment or rock core archives and existing platforms, and which questions would require new infrastructure or sampling investments?

The final report will address the following:

1. Identify novel opportunities regarding ocean-related, use-inspired, solutions-oriented research and innovation. This assessment will include specific examples of opportunities for the Division to make substantial contributions to and develop collaborative and complementary research efforts with NSF's Directorate for Technology, Innovation, and Partnerships (TIP).
2. Identify opportunities and strategies to promote innovative multidisciplinary and multi-sectoral approaches to address complex science challenges arising from the intersection of natural processes, societal needs, and human-driven environmental change. This will include strategies for training the next generation of ocean scientists and incorporating the principles of diversity, equity, inclusion, environmental justice, and access into these scientific endeavors.

3. Develop a concise portfolio of compelling, high-priority, scientific questions that have the potential to transform scientific knowledge of the ocean and the critical role of the ocean in the Earth system. Identification of the scientific questions will update the priorities identified in *Sea Change: Decadal Survey of Ocean Sciences 2015–2025*, drawing from recent reports and community input, including recent National Academies reports and activities such as the U.S. National Committee for the UN Decade of Ocean Science for Sustainable Development. The selection may be based on timeliness, societal benefits, technological advances, or other criteria as identified by the committee.
4. Identify the research infrastructure needed to advance the high-priority ocean science research questions identified in Task # 3, including an assessment of current facilities and the potential for future investments and development of new technologies to meet the needs of the research community. The assessment will include the committee's perspectives on the relative need for continued funding of specific infrastructure and mechanisms to evaluate the contributions of major infrastructure to the research enterprise.
5. Develop a framework that OCE can apply to leverage and complement the capabilities, expertise, and strategic plans of its partners (other NSF units, federal agencies, private sector—such as ocean industries and foundations, and international organizations). The framework will include approaches to encourage greater collaboration and maximize shared use of research assets and data.
6. In undertaking these tasks, the committee will engage the ocean science community and other relevant fields to gather ideas and develop recommendations informed by broader community perspectives. The final report will include assessment of challenges and identification of metrics for progress in achieving the vision of the decadal survey. The DSOS 2025-2035 committee will address these tasks within the context of the current OCE budget while identifying aspirational goals that NSF could implement with growth in the OCE budget over the decade.

# Appendix B

# Committee Meeting Agenda and Participant List

August 2–3, 2023
COMMITTEE ON 2025–2035 DECADAL SURVEY OF OCEAN SCIENCES
The National Academy of Sciences Building, 2101 Constitution Ave NW
Washington, DC 20001

**AUGUST 2, 2023**
**ROOM 120**

| | |
|---|---|
| 9:00–9:15 | Welcome, Introductions, and Meeting Goals<br>**Tuba Özkan-Haller**, Committee Co-Chair<br>**Jim Yoder**, Committee Co-Chair |
| 9:15–10:00 | Overview of the IODP Program<br>**Mitch Malone**, Texas A&M University |
| 10:00–10:15 | Briefing from July IODP Town Hall<br>**Jim McManus**, OCE NSF |
| 10:15–11:00 | Briefing from the 2050 Framework of Scientific Ocean Drilling<br>**Anthony Koppers**, Oregon State University |
| 11:00–11:30 | Break |
| 11:30–12:30 | Open Discussion and Panel on the Future of Scientific Ocean Drilling<br>**Anthony Koppers**, Oregon State University<br>**Adriane Lam**, Binghamton University<br>**Patrick Fulton**, Cornell University<br>**Kathie Marsaglia**, California State University, Northridge<br>**Jim Yoder**, Committee Co-Chair |

| | |
|---|---|
| 12:30–1:00 | Lunch |
| 1:00–1:30 | A Future Vision from LEAPS Report<br>**Larry Krissek**, Ohio State University |
| 1:30–2:00 | Benefits to Overlapping, Cross-disciplinary Ocean and Paleo-Ocean Research Priorities<br>**Daniel Sigman**, Princeton University |
| 2:00–2:45 | Future Perspectives on Scientific Ocean Drilling<br>**Masako Tominaga**, Woods Hole Oceanographic Institution<br>**Chijun Sun**, National Center for Atmospheric Research<br>**Adriane Lam**, Binghamton University<br>**Chris Lowery**, University of Texas<br>**Allyson Tessin**, Kent State University<br>**Jason Sylvan**, Texas A&M University<br>**Jessica Labonté**, Texas A&M University, Galveston<br>**Brandi Kiel Reese**, University of South Alabama<br>**Patrick Fulton**, Cornell University<br>**Tuba Özkan-Haller**, Committee Co-Chair |
| 2:45–4:00 | Breakout Groups to Discuss Science Priorities |
| 4:00–4:15 | Break |
| 4:15–5:15 | Report Back from Breakout Groups and Discussion and Convergence on Highest Priorities |
| 5:15–5:30 | Closing Remarks and Preparation for Day 2<br>**Tuba Özkan-Haller**, Committee Co-Chair<br>**Jim Yoder**, Committee Co-Chair |
| 5:30–6:30 | Light Dinner |
| 6:30 | Adjourn Public Session |

COMMITTEE ON 2025–2035 DECADAL SURVEY OF OCEAN SCIENCES
The NAS Building, 2101 Constitution Ave NW
Washington, DC 20001

**AUGUST 3, 2023**
**ROOM 120**

| | |
|---|---|
| 9:00–9:30 | Briefing from Science Mission Requirements<br>**Becky Robinson**, University of Rhode Island |
| 9:30–10:15 | Panel Discussion on New Options in Infrastructure<br>**Sean Gulick**, University of Texas<br>**Maureen Walczak**, Oregon State University<br>**Becky Robinson**, University of Rhode Island<br>**Carl Brenner**, U.S. Science Support Program<br>**Rick Murray**, Committee Member |

10:15–10:30 Break

10:30–11:30 Breakout Rooms to Discuss Advantages and Disadvantages of Infrastructure Options

11:30–12:30 Report-out from Working Groups and Discussion

12:30–12:45 Closing Remarks and Next Steps

12:45–1:00 Lunch and Adjourn Public Session

*Invited Guests:*

**Jennifer Biddle,** University of Delaware
**Donna Blackman,** University of California, Santa Cruz
**Stefanie Brachfeld,** Montclair State University
**Carl Brenner,** U.S. Science Support Program
**Steven D'Hondt,** University of Rhode Island
**Patrick Fulton,** Cornell University
**Sean Gulick,** University of Texas
**David Hodell,** University of Cambridge
**Celli Hull,** Yale University
**Minoru Ikehara,** Kochi University
**Fumio Inagaki,** Japanese Agency for Marine-Earth Science and Technology
**Brandi Kiel Reese,** University of South Alabama
**Anthony Koppers,** Oregon State University
**Larry Krissek,** Ohio State University
**Jessica Labontè,** Texas A&M University, Galveston
**Adriane Lam,** Binghampton University
**Chris Lowery,** University Of Texas
**Mitch Malone,** Texas A&M University
**Kathie Marsaglia,** California State University, Northridge
**Charna Meth,** IODP Science Support Office
**Heiko Palike,** University of Bremen
**Becky Robinson,** University Of Rhode Island
**Demian Saffer,** University of Texas
**Daniel Sigman,** Princeton University
**David Smith,** University of Rhode Island
**Chijun Sun,** National Center for Atmospheric Research
**Jason Sylvan,** Texas A&M University
**Allyson Tessin,** Kent State University
**Masako Tominaga,** Woods Hole Oceanographic Institution
**Maureen Walczak,** Oregon State University
**Trevor Williams,** Texas A&M University

***Online Registrants:***

Natsue Abe
Anteneh Abiy
Gary Acton
James Allan
Katherine Allen
Brendan Anderson
Gerald Auer
James Austin
Joeven Austine Calvelo
Simone Baecker Fauth
Barbara Balestra
Chandranath Basak
Keir Becker
Thomas Belgrano
Melissa Berke
Jennifer Biddle
Clara Blättler
Samantha Bombard
Chiara Borrelli
Brian Boston
Samantha Bova
Caleb Boyd
Stefanie Brachfeld
Collin Brandl
Simon Brassell
Julie Brigham-Grette
Henk Brinkhuis
Ashley Burkett
Elizabeth Canuel
Sami Cargill
Joe Carlin
Lindsey Carter
James Carton
Humberto Carvajal-Chitty
Laurel Childress
Beth Christensen
Gail Christeson
Leon Clarke
Steven Clemens
Lisa Clough
Bernard Coakley
Jason Coenen
Rosalind Coggon
John Collins
Cathy Constable
Ann Cook
Jack Cooper
Sharon Cooper
Carol Cotterill
Brian Cowden
Stefano Crivellari
Thomas Cronin
Emily Cunningham
Elaine Daniloff
Catherine Davis
Narelle Maia De Almeida
Anne De Vernal
Laura De Santis
Jeanette Decuba
Penny Demetriades
Letizia Di Bella
Cynthia Dinwiddie
Federica Donda
Isabel Dove
Andrea Dutton
Taylor Elpers
Elva Escobar
Emily Estes
Sandra Fogg
Sean Fowler
Mathew Fox
Allison Franzese
Patricia Fryer
Rina Fukuchi
Heather Furlong
Marlo Garnsworthy
Andrew Gase
Daniel Gaskell
Bill Gilhooly
Maya Gomes
Gomez Gonzalez
Rachael Gray
Elizabeth Griffith
Joan Grimalt
Laura Guertin
Mathis Hain
Lydia Hayes
Lauren Haygood
Shannon Haynes
Karla Heidelberg
Oliverkhiowboong Heng
Tim Herbert
Robert Hershey
Anya Hess
Verena Heuer
Luan Heywood
John Higgins
Sean Higgins
David Hodell
Brianna Hoegler
G. Leon Holloway
Gilbert Hong
Brian Huber
Julie Huber
Susan Humphris
Katie Inderbitzen
John Jamieson
Debadrita Jana
Matthew Jones
Frank Jordan
Sarah Kachovich
Emily Kaiser
Pamela Kempton
Hae-Cheol Kim
Kristin Kimble
Masa Kinoshita
Ravi Kiran Koorapati
Sandra Kirtland Turner
Hiroko Kitajima
Kevin Konrad
Louise Koornneef
Aditya Kumar
Akintunde Kuye
Chris Laabs
Susan Lang
Vera Lawson
R. Mark Leckie
Ricardo Letelier
Amy Leventer
Kuan-Yu Lin
Lorraine Lisiecki
Dan Lizarralde
Guido Lueniger
Min Luo
Annette Lyle
Mitchell Lyle
Susanne M. Straub
Renê Magalhães
Tom Maier
Alberto Malinverno
Michael Manga
Roberta Marinelli
Juan Martin
Susana Martín Lebreiro
Stephen Mattin
Joseph Mayala Nsingi
Andrew McCaig
Robert McKay

Lisa McNeill
Danilo Mero
Kenneth Miller
Mike Miller
Rupert Minnett
Bryce Mitsunaga
Kyaw Moe
Alessandra Montanini
Yuki Morono
Gregory Mountain
Joshu Mountjoy
Bastian Muench
Fawz Naim
Malarkodi Nallamuthu
Lucien Nana Yobo
Jared Nirenberg
Tatsuo Nozaki
Osahon Ogbebor
Javier Ojeda
Beth Orcutt
Jeremy Owens
Tim Parker
Ross Parnell-Turner
Molly Patterson
Erin Peck
Michelle Penkrot
Ligia Perez-Cruz
Lorri Peters
Larry Peterson
Katerina Petronotis
Maya Pincus
Pratigya Polissar
Camilo Ponton
Eirini Poulaki
Mohammed Rabiu
Patrick Rafter
Ronnakrit Rattanasriampaipong
Mark Reagan
Margo Regier
Natascha Riedinger
Ulla Röhl
Diana Roman
Karen Romano Young
Thomas Ronge
Yair Rosenthal
Claire Routledge
Jeffrey Ryan
Ash S
Ninette Sadusky
Sanny Saito
Alessio Sanfilippo
Danielle Santiago Ramos
Sara Satolli
Mitch Schulte
Amelia Shevenell
Brandon Shuck
Weimin Si
Elizabeth Sibert
Lexa Skrivanek
Angela Slagle
Sara Smith
Hiroki Sone
Patricia Standring
Scott Starratt
Andrew Steen
Joann Stock
Brittany Stockmaster
Lowell Stott
Danielle Sumy
Rosalynn Sylvan
Akihiro Tamura
Damon Teagle
Kaustubh Thirumalai
Ellen Thomas
Andy Thompson
Emily Tibbett
Sean Toczko
Kayla Tozier
Benjamin Tutolo
Natalie Umling
Jaime Urrutia Fucugauchi
Vinton Valentine
Ellen Varekamp
Natalia Varela
Helenice Vital
Natalie Vu
Shelby Walker
Yi Wang
Michael Webb
Jody Webster
Kristen Weiss
Julia Wellner
Sophie Westacott
Charles Wheat
Scott White
Alicia Wilson
Gisela Winckler
Adam Woodhouse
Jim Wright
Jonathan Wynn
Weimu Xu
Stacy Yager
Kosei Yamaguchi
Michiko Yamamoto
Hong Yang
Mark Yu
Xiaodong Zhang
Yan Zhang
Lin Zho

# Appendix C

# Committee Biographies

**H. Tuba Özkan-Haller** (*Co-Chair*) is dean and professor at Oregon State University, College of Earth, Ocean, and Atmospheric Sciences. She has spent most of her career at the intersection of physical oceanography, marine geology, and coastal and ocean engineering. Her research focuses on the use of numerical, field, laboratory, and analytical approaches to arrive at a predictive understanding of ocean waves, circulation, and coastal change. Özkan-Haller has also extensively engaged in work to increase diversity and inclusivity in academia. She is the recipient of the Office of Naval Research Young Investigator Award, the Outstanding Faculty Member Award at the University of Michigan, and the Pattullo Award for Excellence in Teaching Award at Oregon State University. Özkan-Haller received a B.S. in civil engineering from Boğaziçi University, Turkey and an M.C.E. and a Ph.D. in civil engineering from the University of Delaware. Özkan-Haller serves on the advisory committee to the Army Corps of Engineers Board on Coastal Engineering Research. She previously served on the Ocean Studies Board, as a member of the Marine and Hydrokinetic Energy Assessment Committee, and as chair of the Committee on Long-Term Coastal Zone Dynamics. In April 2021, she published the article "It's Time to Invest in Curiosity" in the *Fair Observer*.

**James (Jim) A. Yoder** (*Co-Chair*) is dean emeritus at the Woods Hole Oceanographic Institution (WHOI) and professor emeritus at the University of Rhode Island (URI). His first academic position was as a researcher at the Skidaway Institute of Oceanography, participating in an interdisciplinary study of the southeastern U.S. continental shelf. He joined the URI faculty in 1989, where he studied regional to global distributions of phytoplankton biomass and productivity using satellite and aircraft measurements. In 2005, Yoder moved to WHOI, where he served as vice president for academic programs and dean. During his academic career, he held temporary assignments as program officer at the Headquarters of the National Aeronautics and Space Administration and as director of the National Science Foundation's Division of Ocean Sciences. He was selected fellow of The Oceanography Society in 2012 and fellow of the American Association for the Advancement of Science in 2019. Yoder received a B.A. from DePauw University and an M.S. and a Ph.D. in oceanography from the University of Rhode Island. He previously served on the Ocean Studies Board and as a member of the 2013-2015 Committee on the Decadal Survey of Ocean Sciences and as chair of the Committee on Catalyzing Opportunities for Research in the Earth Sciences (CORES): A Decadal Survey for NSF's Division of Earth Sciences. Yoder serves on the Corporation of the Woods Hole Oceanographic Institution.

**Lihini Aluwihare** is professor in marine chemistry and geochemistry at the University of California, San Diego, Scripps Institution of Oceanography. She is a chemical oceanographer who studies the cycling of carbon and nitrogen in the oceans using light isotope tools and organic matter chemical characterization. Her work is focused around developing new analytical tools and research frameworks to read the messages encoded in molecules that maintain microbial life, facilitate ecosystem interactions, and contribute to long-term carbon and nutrient storage. She is also interested in the distribution and cycling of anthropogenic compounds in coastal environments. Her career in academia has been guided by a need to build a community of scholars that adequately represents the interests and experiences of the broader population. Aluwihare received a B.S. in chemistry and philosophy from Mount Holyoke College and a Ph.D. from the Massachusetts Institute of Technology-Woods Hole Oceanographic Institution Joint Program in Oceanography.

**Mona Behl** is associate director of Georgia Sea Grant at the University of Georgia, where she also holds public service and academic appointments. She was formerly a policy fellow with the American Meteorological Society's (AMS's) policy program. Her research interests include assisting local communities in managing the impacts of weather and climate, broadening participation in geosciences, and preparing people for the future of work. Behl co-chairs the Mentoring Physical Oceanography Women to Increase Retention program. She also serves on the associate board of directors for the Earth Science Women's Network and the board of directors for the Institute of Georgia Environmental Leaders. She cofounded the AMS Early Career Leadership Academy and Sea Grant's Community Engaged Internship program. For her leadership in advancing the principles of diversity, equity, inclusion, justice, and access within and beyond the national Sea Grant network, Behl was recognized with Sea Grant Association's President's Award in 2020. She was elected to serve on the councils of The Oceanography Society and the AMS in 2022. Behl received a B.S. and an M.S in physics from Panjab University, India, and a Ph.D. in physical oceanography from Florida State University.

**Mark Behn** is associate professor in the Morrissey College of Arts and Sciences at Boston College. His research investigates the dynamics of Earth deformation in glacial, marine, and terrestrial environments through the use of a wide range of geophysical techniques. These techniques include the development of geodynamic models that relate laboratory-based rheologic and petrologic models to the large-scale behavior of Earth, which are then applied to a spectrum of problems from basic science to societally relevant issues. Behn's research interests include dynamics of faulting, magmatism, and surface processes at midocean ridges and continental rifts; seismic anisotropy and imaging of subasthenospheric mantle flow; evolution of the continental crust; and ice sheet dynamics. He is co-chair of the Geodynamics Focus Research Group for the Community Surface Dynamics Modeling System, co-chair for the SZ4D Modeling Collaboratory for Subduction, and former fellow of the Woods Hole Oceanographic Institution Deep Ocean Exploration Institute. Behn received a B.S. in geology from Bates College and a Ph.D. in marine geophysics from the Massachusetts Institute of Technology-Woods Hole Oceanographic Institution Joint Program. He currently serves on the National Academies' Board on Earth Sciences and Resources' Committee on Solid Earth Geophysics.

**Brad deYoung** is executive director of the Pacific node of the Canadian Integrated Ocean Observing System (CIOOS) and professor emeritus and Robert A. Bartlett professor of oceanography at Memorial University. In addition to participating in national and international observing programs, such as the Overturning in the Subpolar North Atlantic Program, he has also been engaged in developing links to public policy and exploring opportunities to connect science, society, and economy. He served for a decade on the Canadian Fisheries Resource Conservation Council, advising the minister of fisheries and oceans on fisheries policy and management. He is working with ocean gliders to make year-round measurements in the Northwest Atlantic and is helping to develop the CIOOS observing network to support data access and the provision of new information services. DeYoung received a Ph.D. in physical oceanography from the University of British Columbia.

**Carlos Garcia-Quijano** holds a joint appointment as professor in the Department of Sociology and Anthropology and the Department of Marine Affairs at the University of Rhode Island. He has special interest in how human cognition, culture, and society influence the interaction between people and the nonhuman environment, as well as who bears the impacts and responsibility for environmental problems. His research interest is focused on comparative study of cultural aspects of coastal use and dependence to reach more comprehensive understanding of human well-being and adaptations as they relate to the use of coastal environments and resources, including local knowledge, resource management, and adaptations to species translocations. Garcia-Quijano received a B.S. in biology and an M.S. in geology and reef paleoecology from the University of Puerto Rico, and a Ph.D. in ecological and environmental anthropology from the University of Georgia.

**Peter Girguis** is professor of organismic and evolutionary biology and co-director of the Microbial Sciences Initiative at Harvard University. He is also adjunct professor in the Woods Hole Oceanographic Institution's Applied Ocean Physics and Engineering group. Girguis is a microbiologist, biogeochemist, and technologist who studies how animals and microbes in the deep sea influence biogeochemical cycles. He is also known for developing novel "open-design" instruments, such as underwater mass spectrometers and microbial samplers, and strives to make these tools available to the broadest research community with the goal of democratizing science around the world. He was a National Science Foundation RIDGE-2000 distinguished lecturer, a Merck Co. Innovative Research Awardee, a recipient of the 2007 and 2011 Lindbergh Foundation Award for Science & Sustainability, the 2018 Lowell Thomas Award in Marine Science and Technology for groundbreaking advances, and the 2020 Petra Shattuck Award for Distinguished Teaching. He was recently named a Gordon and Betty Moore Foundation investigator. Girguis received a B.S. from the University of California, Los Angeles, and a Ph.D. from the University of California, Santa Barbara, and completed postdoctoral research at the Monterey Bay Aquarium Research Institute. He is a member of the National Oceanic and Atmospheric Administration's Ocean Exploration Advisory Board. He was previously a member of the National Academies of Sciences, Engineering, and Medicine's First Indian-American Frontiers of Science Symposium.

**Leila J. Hamdan** serves as associate vice president for research, coastal operations and director and professor in the School of Ocean Science and Engineering at the University of Southern Mississippi. Her research centers on marine microbial biogeography and exploring natural and human-made features on the seabed that shape coastal to deep-sea ecosystems. She is the lead on the National Science Foundation award for the operation of the future Regional Class Research Vessel *Gilbert R. Mason* and has been chief scientist on 25 oceanographic research expeditions. She is currently president of the Coastal and Estuarine Research Federation. She received the National Oceanographic Partnership Program Excellence in Partnering Award in 2017 for leadership of a team of scientists studying impacts of the *Deepwater Horizon* oil spill. Hamdan received a B.S. in biology from Rowan University of New Jersey and an M.S. and a Ph.D. from George Mason University and completed postdoctoral training at the Naval Research Laboratory.

**Marcia J. Isakson** is director of the Signal and Information Sciences Laboratory at Applied Research Laboratories at the University of Texas at Austin. She has expertise in undersea warfare and signal processing, and her research interests include ocean acoustic propagation and autonomous underwater vehicle sensors. She is a fellow and former president of the Acoustical Society of America. Isakson received her B.S. in engineering physics and mathematics from the United States Military Academy at West Point. Upon graduation, she was awarded a Hertz Foundation fellowship and completed an M.S. and a Ph.D. in physics from the University of Texas at Austin. She is a member of the Ocean Studies Board and a member of the U.S. National Committee for the Decade of Ocean Science for Sustainable Development.

**Jason Link** is senior scientist for ecosystems with the National Oceanic and Atmospheric Administration's National Marine Fisheries Service, leading efforts to support development of ecosystem-based management plans and activities throughout the agency. He has held many past adjunct positions and currently holds an adjunct faculty position at the School for Marine Science and Technology at the University of Massachusetts. His research interests include marine resource–ecosystem modeling methodologies, marine food web topology, globally consistent patterns in ecosystem cumulative biomass distributions, and delineation of ecosystem overfishing thresholds.

Link is a fellow of the American Institute of Fishery Research Biologists, was a Frohlich fellow, holds executive leadership certificates from the Harvard Kennedy School and the Key Program at American University, and has received the Fisheries Society of the British Isles Medal for significant advances in fisheries science and a Department of Commerce Bronze medal. He received his B.S. in biology from Central Michigan University and a Ph.D. in biological sciences from Michigan Technological University.

**Allison Miller** is research portfolio senior manager at Schmidt Ocean Institute. She is responsible for managing and overseeing all research projects undertaken by Schmidt Ocean Institute, including management of grants and contracts, and assuring accessibility of all data that are created and collected as part of those projects. Previously, she facilitated federal partnerships by managing the National Oceanographic Partnership Program, housed at the Consortium for Ocean Leadership. Miller serves as secretary of the Council of The Oceanography Society and serves on the external advisory committee of the University Corporation for Atmospheric Research Community Program. Miller received a B.S. in marine science from Coastal Carolina University and an M.S. in oceanography from Florida State University.

**S. Bradley Moran** is dean of the College of Fisheries and Ocean Sciences at the University of Alaska Fairbanks. Prior to his appointment as dean, he served as acting director of the Obama administration's National Ocean Council, assistant director for ocean sciences in the White House Office of Science and Technology Policy, program director in the Chemical Oceanography Program at the National Science Foundation, and professor of oceanography in the Graduate School of Oceanography at the University of Rhode Island. His principal research interests focus on the application of uranium-series and artificial radionuclides as tracers of marine geochemical processes and fostering economic development partnerships in energy and environmental research, technology, policy, and education. He currently serves as vice president of the international Scientific Committee on Oceanic Research; on the board of directors of the Alaska Ocean Observing System, the Alaska Sea Life Center, and the North Pacific Research Board; and, previously, as board chair and trustee of the Consortium for Ocean Leadership. Moran received a B.Sc. in chemistry from Concordia University and a Ph.D. in oceanography from Dalhousie University. He previously served on the National Academies of Sciences, Engineering, and Medicine Ocean Studies Board and is currently an ex officio member of the Ocean Studies Board as a representative to the Scientific Committee on Oceanic Research, and he is a member of the U.S. National Committee for the Decade of Ocean Science for Sustainable Development.

**Richard W. (Rick) Murray** is deputy director and vice president for science and engineering at Woods Hole Oceanographic Institution. Prior to this role, he was professor of Earth and environment at Boston University from 1992 to 2019, and served as director for the Division of Ocean Sciences at the National Science Foundation from 2015 to 2018. Murray also served as co-chair for the Subcommittee on Ocean Science and Technology, as part of the Office of Science and Technology Policy in the Executive Office of the President, during both the Obama and Trump administrations. His research interests are in marine geochemistry, with an emphasis on sedimentary chemical records of climate change, volcanism, and tropical oceanographic processes, and in the chemistry of the subseafloor biosphere. Murray was involved in the advisory structure for the scientific ocean drilling programs throughout his career. He sailed on six different scientific drilling expeditions (127, 165, 175, 185, 329, and 346), including as co-chief scientist on Expedition 346. Murray is a former councilor of The Oceanography Society and a former member of the board of directors of the American Geophysical Union. Murray received a B.A. in geology from Hamilton College and a Ph.D. in geology and geophysics from the University of California, Berkeley. He also graduated from the Sea Education Association's program at Woods Hole Oceanographic Institution and completed postdoctoral training at the University of Rhode Island's Graduate School of Oceanography. Murray has testified before the U.S. Congress and the Massachusetts State Legislature on issues relating to climate change and ocean observations.

**Stephen R. Palumbi** is Jane and Marshall Steel Professor in Marine Sciences and senior fellow with the Woods Institute for the Environment at Stanford University. He is former director for Hopkins Marine Station at Stanford.

His research interests include the use of molecular genetics techniques to study evolution and change within marine populations. He has contributed to enhancing understanding of speciation patterns in open-ocean systems, providing insights for marine reserve design and refuges for thermally sensitive corals. He received the Peter Benchley Award for Excellence in Science and is an elected member of the National Academy of Sciences, fellow of the California Academy of Sciences, and Pew fellow in marine conservation. Palumbi has published three books on science for the general public, cofounded the microdocumentary series *Short Attention Span Science Theater*, and has appeared in numerous ocean documentaries. He received a B.A. in biology from Johns Hopkins University and a Ph.D. in zoology with a concentration in marine ecology from the University of Washington.

**Ella (Josie) Quintrell** recently retired as founding executive director of the Integrated Ocean Observing System (IOOS) Association and now serves as senior advisor to the organization. The IOOS Association works with the IOOS Regional Associations to design and operate coastal observing systems to collect, integrate, and produce information for users. Her research interests focus on operational observing issues related to harmful algal blooms, cloud computing, and coastal management. She previously served on the board of the Consortium for Ocean Leadership, the National Estuarine Research Reserve System Science Advisory Committee, and the National Marine Association of Marine Laboratories. Quintrell received a B.A. in biology from Colby College and an M.R.P. in environmental planning from Cornell University.

**Yoshimi (Shimi) Rii** is assistant specialist at Hawai'i Institute of Marine Biology and also serves as research coordinator for the He'eia National Estuarine Research Reserve within the National Oceanic and Atmospheric Administration's Office for Coastal Management. Her expertise is in marine ecology, with phytoplankton and nutrient dynamics in coastal and open-ocean environments. She utilizes multiple tools spanning geochemical tracers, biomarkers, and genomic approaches to examine metabolic activities and biodiversity of microbial eukaryotes. Her work encompasses research at the watershed to coastal ocean scale, examining the continuum and diversity of ecologically important species and processes in a diverse environmental setting. She is passionate about increasing diversity and equity, and promoting the inclusion of multiple ways of knowing, including Indigenous knowledge, within conventional science within the academy. Rii received a B.S. in marine biology and English from the University of California, Los Angeles, and an M.S. and a Ph.D. in biological oceanography from the University of Hawai'i at Mānoa.

**Kristen St. John** is professor of geology at James Madison University. Her research focuses on marine sediment records of past climate change and on teaching and learning in the geosciences. As an active researcher in the International Ocean Discovery Program (IODP) and legacy programs, she participated as a marine sedimentologist for several expeditions; is serving as co-chief scientist for Expedition 403, scheduled in summer 2024 on the *JOIDES Resolution*; and is a coproponent on an active IODP proposal for scientific drilling in the Arctic Ocean for consideration by the European Consortium for Ocean Research Drilling. She is president of the American Geophysical Union Education Section and fellow of the Geological Society of America, and was editor-in-chief of the *Journal of Geoscience Education* from 2012 to 2017. St. John received an M.S. and a Ph.D. in geoscience from Ohio State University. She currently serves on the Polar Research Board and previously chaired the Workshop on Tipping Points, Cascading Impacts, and Interacting Risks in the Earth System and was a member of the Committee on Advancing a Systems Approach to Studying the Earth: A Strategy for the National Science Foundation. She served on the workshop steering committee and as co-author of the report for the NEXT: Scientific Ocean Drilling Beyond 2023.

**Samuel K. (Kersey) Sturdivant** is principal scientist at INSPIRE Environmental; adjunct assistant professor at Duke University; and cofounder of Oceanography for Everyone, an open-source effort to develop low-cost oceanographic hardware. His research interests center broadly around the effects of human disturbance on the seafloor and development of marine technology to enhance human understanding of the ocean. Sturdivant gave a TED talk, "Visualize the Seafloor" and published a comprehensive step-by-step guide on how to get into graduate school in the sciences with Cambridge University Press. Sturdivant received a B.S. in environmental science from the University of Maryland Eastern Shore and a Ph.D. in marine science from the College of William & Mary's Virginia Institute of Marine Science.

**Ajit Subramaniam** is research professor at the Lamont-Doherty Earth Observatory (LDEO) of Columbia University. He has served as program director for the Marine Microbiology Initiative at the Gordon and Betty Moore Foundation and as program director in the Biological Oceanography Program at the National Science Foundation while on leave from LDEO. Subramaniam is a microbial oceanographer with expertise in biogeochemical cycles, remote sensing, bio-optics, and phytoplankton physiology. His research interests focus on advancing the ability to observe the ocean and expand understanding on how the marine ecosystem works and can be managed. He was awarded a Mercator fellowship by the University of Rostock and the Baltic Sea Research Institute, Germany, in 2017 and the Climate and Life fellowship at Lamont-Doherty Earth Observatory in 2021. Subramaniam received a B.S. in physics from the American College in India, and an M.S. in marine environmental science and a Ph.D. in coastal oceanography from the State University of New York at Stony Brook.

**Maya Tolstoy** is Maggie Walker Dean of the University of Washington College of the Environment. Prior to this role, Tolstoy she was professor at Columbia University's Department of Earth and Environmental Sciences at Lamont-Doherty Earth Observatory, and previously served as interim executive vice president and dean of the faculty of Arts and Sciences at Columbia. She is a marine geophysicist specializing in seafloor earthquakes and volcanoes. Over her more than 30-year career as a researcher, professor, and administrator, Tolstoy has dedicated herself to furthering understanding of the fundamental processes of the planet and advancing diversity, equity, and inclusion in academia. Tolstoy was awarded the Wings Worldquest Sea Award honoring women in exploration and was a finalist for the National Aeronautics and Space Administration's 2009 Astronaut selection. Tolstoy received a B.S. in geophysics from the University of Edinburgh and a Ph.D. from Scripps Institution of Oceanography at the University of California, San Diego. She previously served on the National Academies' Board on Earth Sciences and Resources' Committee on Solid Earth Geophysics.

**Shannon Valley** is climate advisor with Vistant, contracted to the U.S. Agency for International Development's Center for Climate Positive Development. Prior to this role, she served as legislative liaison at Headquarters for the National Aeronautics and Space Administration (NASA) and as policy assistant in the White House Domestic Policy Council, and was appointed to the 2020 Presidential Transition NASA Agency Review team. Her past research focused on Atlantic meridional overturning circulation variability and its relation to abrupt and long-term climate change. She is a recipient of the NASA Individual Special Act Award, the NASA Exceptional Achievement Medal, and the National Science Foundation Graduate Research Fellowship. Valley received a B.A. in political science and international studies from Northwestern University and an M.S. and a Ph.D. in Earth and atmospheric science from the Georgia Institute of Technology. She completed postdoctoral research in paleoceanography and coastal geochemistry, working with marsh sediment cores at Woods Hole Oceanographic Institution.

**James Zachos** is distinguished professor of Earth and planetary sciences and Ida Benson Lynn Chair of Ocean Health at the University of California, Santa Cruz. His research focuses on resolving aspects of the ocean, climate, and carbon-cycle dynamics of the last 65 million years, addressing issues ranging from the causes of extreme greenhouse warming and ocean acidification to the onset of Antarctic glaciation. His past research included participation in the scientific ocean drilling program. He is a member of both the American Academy of Arts and Sciences and the National Academy of Sciences and is a fellow of the American Geophysical Union (AGU), the Geological Society of America, and the American Association for the Advancement of Science. He is also a recipient of the AGU Emiliani Award, the European Geophysical Union Milutin Milankovic Medal, and the BBVA Frontiers of Knowledge award. Zachos received a B.S. in geology/economics from the State University of New York at Oneonta, an M.S. in geology from the University of South Carolina, and a Ph.D. from the Graduate School of Oceanography at the University of Rhode Island. He completed a postdoctoral fellowship at the University of Michigan.